PHILIP'S MONTH-E

D0263534

STARGAZING 2017

THE GUIDE TO THE NORTHERN NIGHT SKY

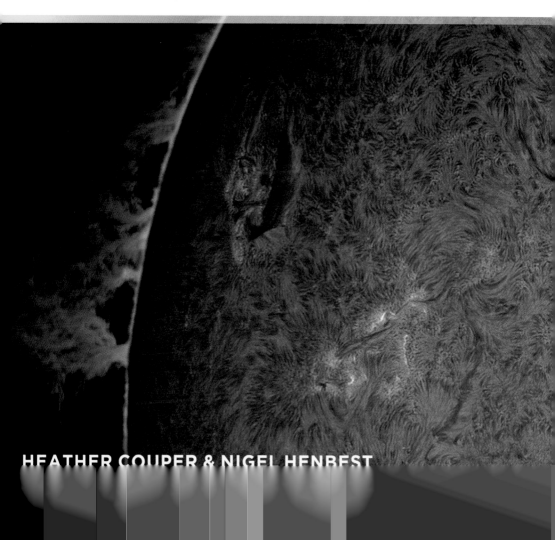

HEATHER COUPER & NIGEL HENBEST

www.philipsastronomy.com
www.philips-maps.co.uk

HEATHER COUPER and NIGEL HENBEST are internationally recognized writers and broadcasters on astronomy, space and science. They have written more than 40 books and over 1000 articles, and are the founders of an independent TV production company specializing in factual and scientific programming.

Heather is a past President of both the British Astronomical Association and the Society for Popular Astronomy. She is a Fellow of the Royal Astronomical Society, a Fellow of the Institute of Physics and a former Millennium Commissioner, for which she was awarded the CBE in 2007. Nigel has been Astronomy Consultant to *New Scientist* magazine, Editor of the *Journal of the British Astronomical Association* and Media Consultant to the Royal Greenwich Observatory.

Heather Couper

Published in Great Britain in 2016 by Philip's,
a division of Octopus Publishing Group Limited
(www.octopusbooks.co.uk)
Carmelite House, 50 Victoria Embankment,
London EC4Y 0DZ
An Hachette UK Company (www.hachette.co.uk)

TEXT
Heather Couper and Nigel Henbest (pages 4–53)
Robin Scagell (pages 61–64)
Philip's (pages 1–3, 54–60)

Copyright © 2016 Heather Couper and Nigel Henbest
(pages 4–53)

Copyright © 2016 Philip's (pages 1–3, 54–64)

ISBN 978–1–84907–425–4

Nigel Henbest

Printed in China

Title page: *Solar prominence*
(Pete Williamson/Galaxy)

ACKNOWLEDGEMENTS

All star maps by Wil Tirion/Philip's, with extra annotation by Philip's. Artworks © Philip's.

All photographs courtesy of Galaxy Picture Library:
David Arditti *36*;
Stuart Atkinson *63 (bottom)*;
Eddie Guscott *16*;
Pete Lawrence *40*;
Meade *63 (top)*;
Optical Vision *62*;
Damian Peach *12, 20, 44*;
Robin Scagell *32, 48, 52, 64*;
Ian Sharp *8, 24*;
Pete Williamson *1, 28*.

CONTENTS

Welcome to the world of stargazing! Within these pages you'll find a complete guide to what's happening in the night sky throughout 2017. If you're new to astronomy – maybe you've been inspired by a shooting star, a supermoon or the Northern Lights – join the Celestial Club! For seasoned observers, there's plenty here to keep you agog all year round.

With the 12 monthly star charts, you can find your way around the sky on any night in the year. Impress your friends by identifying celestial sights, from the brightest planets to some pretty obscure constellations. And we've included updates on everything that's new in 2017, from comets to shooting stars to eclipses.

THE MONTHLY CHARTS

A reliable map is just as essential for exploring the heavens as it is for visiting a foreign country. For each month, we provide two **star charts**, showing the view looking north and south. To keep the maps uncluttered, we've plotted about 200 of the brighter stars (down to third magnitude), which means you can pick out the main star patterns – the constellations. (In total, over 6000 stars are visible to the unaided eye!) We also show the ecliptic: the apparent path of the Sun in the sky, which is closely followed by the Moon and planets. You can use these charts throughout most of Europe, North America and northern Asia – between 40° and 60° north – though our detailed notes apply specifically to the UK and Ireland.

USING THE STAR CHARTS

Start by working out your compass points. South is where the Sun is highest in the sky during the day; east is roughly where the Sun rises, and west where it sets. You can find north by locating the Pole Star – Polaris – at night by using the stars of the Plough (see March).

The left-hand chart shows your view to the north. Most of the stars here are visible all year: these circumpolar constellations wheel around Polaris as the seasons progress. Your view to the south appears in the right-hand chart; it changes much more as the Earth orbits the Sun. Leo's prominent 'Sickle' is high in the spring skies. Summer is dominated by the bright trio of Vega, Deneb and Altair. Autumn's familiar marker is the Square of Pegasus; while the stars of Orion rule the winter sky.

During the night, our perspective on the sky also alters as the Earth spins around, making the stars and planets appear to rise in the east and set in the west. The charts depict the sky in the late evening (the exact times are noted in the captions). As a rule of thumb, if you are observing two hours later, then the following month's map will be a better guide to the stars on view – though beware: the Moon and planets won't be in the right place!

THE MOON, PLANETS AND SPECIAL EVENTS

Our charts also highlight the **planets** above the horizon in the late evening. However, you may be able to spot other planets on the same night, which have either set by the time shown on the chart, or haven't risen yet. Turn to our monthly **Planets on View** notes, which describe what can be seen throughout the hours of darkness.

The position of the Full Moon is plotted each month, and also the **Moon's position** at three-day intervals before and after. The adjacent table has detailed information on the **Moon's phases**. If there's a **meteor shower** in the month, we mark the radiant – the position from which the meteors appear to stream – and describe it more fully in the **Special Events** section. Here you'll also find information on close pairings of the planets, times of the equinoxes and solstices, and – most exciting – **eclipses** of the Moon and Sun. We've also shown the track of any **comets** known at the time of writing; though we can't guide you to a comet found after the book has been printed! Nor can we predict the notoriously fickle appearance of aurorae.

Each month, we examine one particularly interesting **object**: a planet perhaps, or a star or a galaxy. We also feature a spectacular **picture** – taken by a backyard amateur based in Britain – and describe how the image was captured. And we explore a fascinating and often newsworthy **topic**, ranging from star colours to a spacecraft smashing into Saturn.

There's a full annual overview of events in the **Solar System Almanac** on pages 54–57, along with diagrams explaining the motion of the planets and the celestial choreography that leads to an eclipse. We also give details about the magnitude scale, light years and the separation of objects in the sky.

GETTING IN DEEPER

For each month, there's a practical **observing tip** to help you explore the sky with the naked eye, binoculars or a telescope. If you're after 'faint fuzzies' too dim to appear on the charts, turn to the list of **recommended deep-sky objects** (pages 58–60). The adjacent table of 'limiting magnitude' indicates which objects are visible with your equipment.

For a round-up of what's new in **observing technology**, check out pages 61–64, where equipment expert Robin Scagell offers advice on portable equipment you can use to roam the sky while on holiday.

Finally, on cloudy nights or during the day, join in online projects at the cutting edge of astronomy, by following the links in our **Citizen Science** boxes – from probing sunspots to finding a new pulsar.

Happy stargazing!

If ever there was a time to see A-list stars strutting their stuff, it's this month. Find your way around the sky with the help of the stunning winter stars, led by **Betelgeuse** and **Rigel** in the magnificent hunter **Orion**. Nearby, you'll find **Aldebaran**, the bright red eye of **Taurus** (the Bull); **Capella**, crowning **Auriga** (the Charioteer); **Castor** and **Pollux**, the celestial twins in **Gemini**; and glorious **Sirius**, in **Canis Major** (the Great Dog). This year, these celestial jewels are joined by dazzling Venus – a brilliant lantern in the dusk sky, ushering in the New Year.

▼ The sky at 10 pm in mid-January, with Moon positions at three-day intervals either side of Full Moon. The star positions are also correct for 11 pm at

JANUARY'S CONSTELLATION

Spectacular **Orion** is one of the rare star groupings that looks like its namesake – a giant hunter with a sword below his belt, wielding a club above his head.

The seven main stars of this brilliant constellation lie in the 'top 70' brightest stars in the sky. Despite its distinctive shape, most of these stars are not closely associated with each other – they simply line up, one behind the other.

Closest – at 250 light years – is the star forming the hunter's right shoulder, **Bellatrix**. And an interesting puzzle here! A reader has pointed out that if Orion is facing us, Bellatrix should be his *left* shoulder. But we've scoured the mythological engravings, and these show Orion is just as frequently depicted with his *back* to us – so 'right' would be correct.

Next in the hierachy of Orion's superstars is blood-red **Betelgeuse**. It lies at the top left of the constellation, and is 640 light years away. The star is a thousand times larger than our Sun, and its fate will be to explode as a supernova.

The constellation's brightest star, blue-white **Rigel** (Orion's foot), is a vigorous young star more than twice as hot as our Sun, and 125,000 times more luminous. Rigel lies around 860 light years from us.

the beginning of January, and 9 pm at the end of the month. The planets move slightly relative to the stars during the month.

Saiph, marking the hunter's other foot, is 650 light years distant The two outer stars of the belt, **Alnitak** (left) and **Mintaka** (right) lie 700 and 690 light years away, respectively.

We travel 1300 light years from home to reach the middle star of the belt, **Alnilam**. And at the same distance, we find the stars of the 'sword' hanging below the belt – the lair of the great **Orion Nebula** – an enormous star-forming region 24 light years across (see this month's Picture).

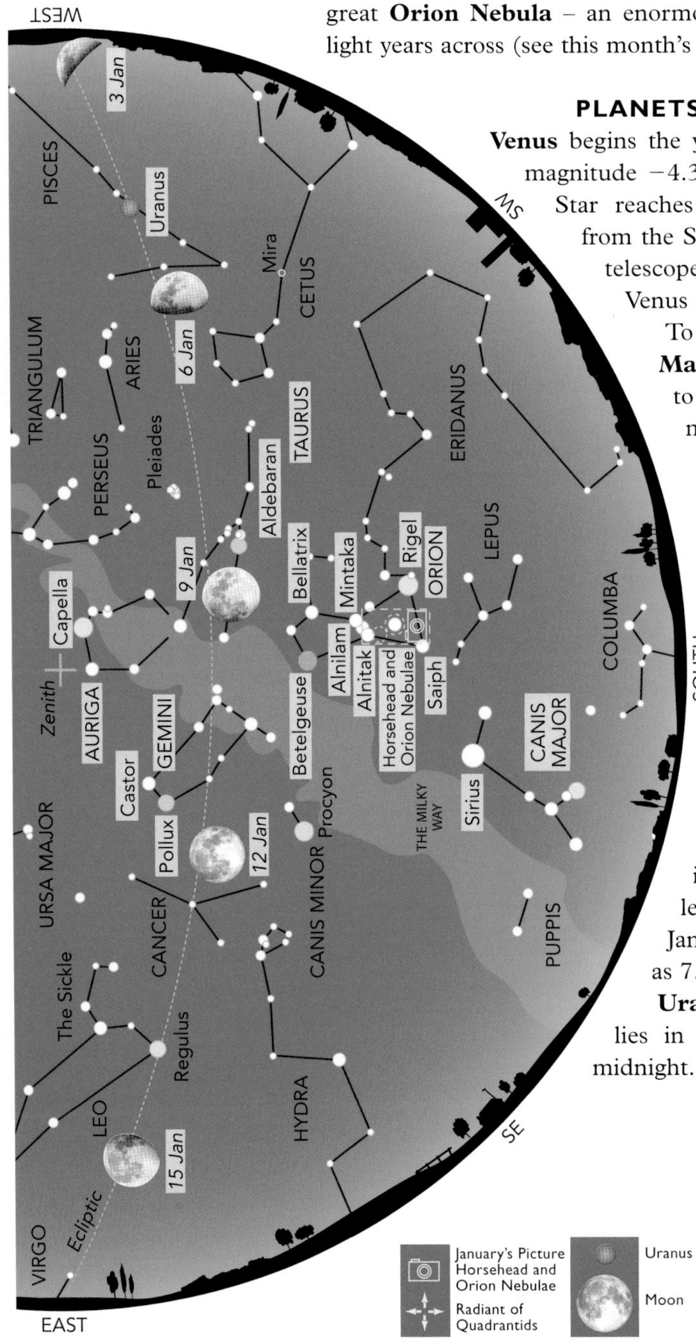

PLANETS ON VIEW

Venus begins the year in a blaze of glory, at magnitude −4.3. The magnificent Evening Star reaches its maximum elongation from the Sun on **12 January**; a small telescope reveals its globe is half-lit. Venus sets around 8.45 pm.

To its upper left you'll find **Mars**, moving from Aquarius to Pisces during January. At magnitude +1.0, the Red Planet is 100 times fainter than Venus, and slips below the horizon at 9 pm.

There are two easy chances this month to spot faint **Neptune** (magnitude +7.9) – also in Aquarius – though you'll need binoculars to see it. On **1 January**, Neptune lies only 20 arc minutes to the lower right of Mars; on **12 January**, it's a similar distance to the left of Venus. By the end of January, Neptune sets as early as 7.30 pm.

Uranus, at magnitude +5.8, lies in Pisces and sets just after midnight.

January's Picture
Horsehead and
Orion Nebulae

Radiant of
Quadrantids

Uranus

Moon

MOON		
Date	Time	Phase
5	7.47 pm	First Quarter
12	11.34 am	Full Moon
19	10.14 pm	Last Quarter
28	0.07 am	New Moon

Giant **Jupiter** rises about midnight, shining brightly at magnitude -1.9 near Virgo's brightest star, Spica.

In mid-January, we have a pair of planets low in the south-east before dawn. The brighter is **Mercury**, which shines at magnitude -0.1 when it reaches greatest western elongation on **19 January**. To its upper right is **Saturn** (magnitude $+0.6$), rising around 6 am in Ophiuchus.

MOON

You'll find a narrow crescent Moon near Venus on **1 January**; between Venus and Mars on **2 January**; and to the upper left of Mars on **3 January**. The star next to the Moon on **9 January** is Aldebaran. The Moon passes Regulus on the night of **14/15 January**. On the night of **18/19 January**, the Moon lies above Jupiter and Spica. Saturn is below the waning crescent Moon on the morning of **24 January**. Back in the evening sky, the crescent Moon revisits Venus and Mars on **31 January**.

SPECIAL EVENTS

On **4 January**, at 2.18 pm, the Earth is at perihelion, its closest point to the Sun – a 'mere' 147 million kilometres away.

The night of **3/4 January** sees the maximum of the **Quadrantid** meteor shower, tiny particles of dust shed by the old comet 2003 EH_1 that burn up – often in a blue or yellow streak – as they enter the Earth's atmosphere. Best seen after the Moon has set at 11 pm.

ESA's BepiColombo – a pioneering mission to Mercury – should launch towards the end of January (see this month's Topic).

JANUARY'S OBJECT

Venus – the planet of love – is resplendent in our evening skies this month. So brilliant and beautiful, she can even cast a shadow in a really dark, transparent sky. Her purity and lantern-like luminosity are beguiling – but looks are deceptive. Earth's twin in size, Venus could hardly be more different from our warm, wet world. The reason for the planet's brilliance is the highly reflective clouds that cloak its surface: probe under these palls of sulphuric acid hanging in an atmosphere of carbon dioxide, and you find a planet out of hell. Volcanoes are to blame. They have created a runaway greenhouse effect that has made Venus the hottest and most poisonous planet in the Solar System. At 460°C, this world is way off the top end of your oven's temperature dial. The pressure at its surface is around 90 Earth-atmospheres. So, if you were to visit Venus, you'd be simultaneously roasted, crushed, corroded and suffocated!

JANUARY'S PICTURE

A spectacular view of star-forming regions in glorious Orion. The **Orion Nebula** (lower right) is easily visible to the unaided eye away from light pollution. The nebula complex above is home to the iconic dark **Horsehead Nebula**, which looks like a cosmic chesspiece. It's made of black interstellar dust – 'soot' from dying stars – which conceals the fledgling stars that are being born inside. The red glow in this image comes from hydrogen gas, heated by fiery newborn stars.

◄ A four-hour exposure of part of Orion, from the star Alnitak and the neighbouring Horsehead Nebula (top) to the Orion Nebula. It was taken using a modified Canon 6D and a Williams Optics WO71 f/4.9 APO refractor. The image was captured by Ian Sharp, who was in Rodalquilar, Spain.

JANUARY'S TOPIC
Mercury

Tiny **Mercury** – the planet closest to the Sun – is putting on an appearance in our morning skies this month. It's currently being surveyed by NASA's Messenger space probe; but now the Europeans are putting in a rival bid. On 27 January, ESA's BepiColombo spacecraft blasts off towards this enigmatic world. The probe is named after the late Guiseppe (Bepi) Colombo, a scientist, engineer and mathematician at Italy's Padua University.

Built as two satellites, BepiColombo will reach Mercury – after a roundabout circuit of the Solar System – in 2024. The Mercury Magnetic Orbiter will separate from the Mercury Orbiter when they arrive.

The aims? To investigate the planet's geology – including its wrinkled, warped surface; its enormous numbers of craters; its composition (including the nature of Mercury's huge metallic core); its magnetic field; and the origin and evolution of a world so close to its parent star.

Venus is even more brilliant this month, joined by the faint comet Encke. The winter star patterns are drifting westwards, as a result of our annual orbit around the Sun. Imagine: you're whirling round on a fairground carousel, and looking out around you. At times you spot the ghost train; sometimes you see the roller-coaster; and then you swing past the candy-floss stall. So it is with the sky: as we circle our local star, we get to see different stars and constellations with the changing seasons.

▼ The sky at 10 pm in mid-February, with Moon positions at three-day intervals either side of Full Moon. The star positions are also correct for

FEBRUARY'S CONSTELLATION

You can't ignore **Gemini** in February. The constellation is crowned by the bright stars Castor and Pollux, representing the heads of a pair of twins, with their bodies running in parallel lines of stars. In legend Castor and Pollux were conceived by the princess Leda on the night she married the king of Sparta, the father of mortal Castor. But Zeus also invaded the marital suite, disguised as a swan, and fathered the immortal Pollux. The pair were so devoted that Zeus placed them together for eternity among the stars.

Castor is an amazing star: a family of six. Even through a small telescope, you can see that Castor is a double star. Both of these stars are themselves double (although you need special equipment to detect this). Then there's another outlying star, visible through a telescope, which also turns out to be double.

Pollux – slightly brighter than Castor – boasts a huge planet: a mighty world bigger than Jupiter.

The constellation contains a pretty star cluster, **M35**. Even at a distance of 2800 light years, it's visible to the unaided eye, and a fine sight when seen through binoculars or a small telescope.

WEST

CETUS
PISCES
ANDROMEDA
TRIANGULUM
ARIES
Ecliptic
Pleiades
PEGASUS
CASSIOPEIA
Algol
PERSEUS
THE MILKY WAY
Capella
AURIGA
Deneb
CEPHEUS
NORTH
CYGNUS
Polaris
Zenith
URSA MINOR
Kochab
URSA MAJOR
LYRA
Vega
DRACO
The Plough
CANES VENATICI
HERCULES
CORONA BOREALIS
NE
BOÖTES
Arcturus
VIRGO

EAST

II pm at the beginning of February, and 9 pm at the end of the month. The planets move slightly relative to the stars during the month.

PLANETS ON VIEW

This month, glorious **Venus** sets four hours after the Sun, and reaches its maximum brightness: magnitude −4.5. Later in February (when the Moon is out of the way) check in a dark location for shadows cast by the Evening Star. A small telescope shows Venus shrinking to a narrow crescent as it speeds towards the Earth.

The 'star' to the upper left of Venus is **Mars**; but it's no match for its gaudy sister, at magnitude +1.2. Setting at 9.30 pm, the Red Planet inhabits Pisces.

Uranus also lies in Pisces, and sets around 10 pm. Use Mars to locate this distant world on **26 February**, when faint Uranus (magnitude +5.9) lies just half a degree to the left of the Red Planet.

Over in the east, **Jupiter** is rising about 10.30 pm, shining at a brilliant magnitude −2.1 in Virgo: the star to the lower right is Spica.

Finally, **Saturn** (magnitude +0.6) rises around 4 am, in Ophiuchus.

Mercury and **Neptune** lie too close to the Sun to be easily visible this month.

MOON

On **2 February**, the crescent Moon forms a line with Mars and Venus to its lower right. As the sky grows dark on **5 February**, you'll find the Moon in front of the Hyades star cluster; later in the evening, it grazes past Aldebaran. The Full Moon lies near Regulus on **11 February**. Jupiter is the brilliant object to the right of the Moon on **15 February**, with

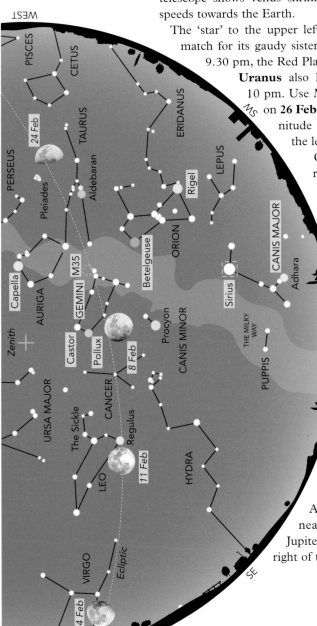

MOON		
Date	**Time**	**Phase**
4	4.19 am	First Quarter
11	0.33 am	Full Moon
18	7.33 pm	Last Quarter
26	2.58 pm	New Moon

February's Object
Sirius

Moon

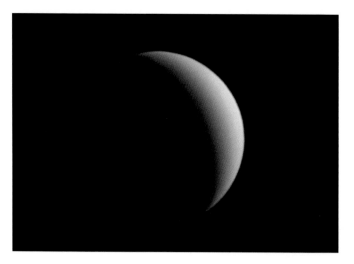

◄ Damian Peach captured Venus from Southgate, London, on 21 February 2009, from a video sequence made using a Celestron C14 355 mm telescope, and a SKYnyx 2.0M camera. He took a sequence of 3000 frames to produce this single image.

Spica nearby. The crescent Moon lies to the left of Saturn on the morning of **21 February**. You may spot a thin waxing crescent Moon on **28 February**, well to the lower left of Venus.

SPECIAL EVENTS

During the second half the month, use binoculars to check out Comet Encke, at its closest approach in several years. You'll find the faint fuzz-ball (magnitude +7) about 10° to the lower right of Venus.

On **26 February**, an annular eclipse of the Sun is visible along a narrow band stretching from southern Chile and Argentina, across the South Atlantic Ocean, to Angola. The Moon appears a little smaller than the Sun, so you'll see a thin ring (annulus) of the Sun's disc around the Moon's silhouette. People in southern regions of South America and western Africa will experience a partial solar eclipse. (More details at eclipse.gsfc.nasa.gov/solar.html)

FEBRUARY'S OBJECT

Sirius – low in the south – is lording it over the night skies this month. At magnitude −1.47, Sirius is the brightest star in the sky, but it's not particularly luminous: it just happens to lie nearby on the cosmic scale. Just 8.6 light years away, Sirius is only twice as distant as our nearest neighbours, the Alpha Centauri trio in the southern hemisphere.

Popularly known as the 'Dog Star', its name derives from the ancient Greek Seirios, which means 'scorcher'. The Greeks also believed that, as Sirius rose with the Sun in late summer, its extra heat heralded the hot and humid 'dog days', when everything slowed down and dogs become lethargic.

To the Egyptians, the summer appearance of Sirius was good news. The annual 'heliacal rising' (when Sirius rose just before

Venus is a real treat this
month. And – if you can
get to view it through
a small telescope – you
have the rare chance
to see another planet
illuminated as a crescent.
But don't wait for the sky
to get totally dark. Seen
against a black sky, the
cloud-wreathed world
is so brilliant it's difficult
to make out any details.
You're best off viewing
Venus soon after the Sun
has set, when the Evening
Star first becomes visible in
the twilight glow. Through
a telescope, the planet
then appears less dazzling
against a pale blue sky.

the Sun) told them that the Nile was about to flood – a good harbinger for a bumper harvest.

Sirius is the brightest star in **Canis Major**, one of Orion's two hunting dogs. Boasting a temperature of about 10,000°C, Sirius is twice as heavy as the Sun. And it's a relatively young star: just 230 million years old, as compared to the Sun's venerable 4,600 million years.

In 1844, Friedrich Bessel noticed that Sirius was being pulled out of place by a faint companion star. This tiny sibling, nicknamed 'the Pup', was once a star much heavier than Sirius, but it has now puffed off its atmosphere to reveal the central nuclear reactor. The same weight as the Sun, and yet the size of the Earth, the Pup is a dense white dwarf, with a searing surface temperature of 25,000°C and considerable gravitational powers. But it's on the road to oblivion. With no nuclear reactions to keep it alive, the Pup will cool to become a dead, black globe. Just wait 2 billion years ....

To spot the Pup – which is 10,000 times fainter than Sirius at magnitude +8.4 – you'll need a 150 mm telescope, viewing through a very steady atmosphere.

FEBRUARY'S PICTURE

Planet **Venus** is the queen of the heavens this month. Earth's twin in size, it's wreathed in thick clouds, as you can see in this image. Only spacecraft carrying powerful radar can penetrate its thick veils, to probe in detail the world's hidden volcanic surface.

FEBRUARY'S TOPIC
Star colours

We think of stars as being white, but check out the glorious gems of midwinter: baleful red **Betelgeuse**, yellow **Capella**, and scintillating blue-white **Rigel**.

These colours (best seen in binoculars) are like a cosmic thermometer, allowing us to take a star's temperature. The hottest stars are blue-white. White stars come next; then yellow, orange and red.

For instance, Betelgeuse's ruddy hue is a sure sign that it's a cool star. Its surface temperature is just 3400°C, as compared to 5500°C for our yellow Sun. This coolish star glows only dully, like an expiring log fire: but its huge size – 1000 times wider than the Sun – means its total luminosity is very high. That's why astronomers call it a 'red giant'.

Capella is hotter, so it has a yellow aura, like the Sun. And Rigel is near the top end of the stellar temperature scale, its blue-white surface at an incandescent 12,000°C. This super-hot star shines 125,000 times more brilliantly than our Sun.

Plenty of reasons to celebrate this month. We have two Evening Stars in the sky – and a simultaneous Morning Star! Plus – the spring has sprung: on **20 March**, we celebrate the Vernal Equinox, when day becomes longer than the night; and British Summer Time starts on **26 March**.

▼ *The sky at 10 pm in mid-March, with Moon positions at three-day intervals either side of Full Moon. The star positions are also correct for 11 pm at*

MARCH'S CONSTELLATION

Ursa Major (the Great Bear) is an internationally favourite star pattern. In Britain, its seven brightest stars are called **'the Plough'**. Most people today have never seen an old-fashioned horse-drawn plough, though, and we've found children naming this star pattern 'the saucepan.' In North America, it's known as 'the Big Dipper'.

The Plough is the first constellation that most people get to know. There are two reasons. First, it's always on view in the northern hemisphere. And second, the two end stars of the 'bowl' of the Plough point directly towards the Pole Star, **Polaris** – which never moves in the heavens, being directly 'over' the North Pole. It's a corner-stone for navigation.

Ursa Major is unusual in a couple of ways. It contains a double star that you can actually split with the naked eye. **Mizar**, the star in the middle of the bear's tail (or the handle of the saucepan) has a fainter companion, **Alcor**. The whole system con-sists of six stars.

And – unlike most constellations – the majority of the stars in the Plough lie at the same distance and were born together. Leaving aside the two end stars, **Dubhe** and **Alkaid**, the others are all moving in the same direction (along with brilliant **Sirius**, which is also a member of the group). Over thousands of years, the shape of the Plough will gradually change, as Dubhe and Alkaid go off on their own paths.

the beginning of March, and 10 pm at the end of the month (after BST begins). The planets move slightly relative to the stars during the month.

PLANETS ON VIEW

Venus starts the month once again as the resplendent Evening Star at magnitude −4.4; but by the end of March it's disappeared from the dusk sky, and has been reborn as the Morning Star. As it passes between Earth and the Sun on **25 March**, Venus appears to fly above the solar north pole; for a few nights before this, we can see Venus both low in the sunset sky, and in the morning rising shortly before the Sun.

Its place in the dusk twilight during the final week of March is taken by **Mercury**, putting on its best evening appearance of the year. The innermost planet is both fainter (magnitude −0.5) and lower in the sky than Venus has been – look near the western horizon at about 8 pm.

Mars is hanging around higher in the evening sky, setting at 9.45 pm. You'll find the Red Planet in Aries, shining at magnitude +1.4.

Uranus (magnitude +5.9) lies in Pisces; Mars is nearby early in March, and Mercury on **25 March**. By month's end, Uranus has dropped into the twilight glow.

Giant planet **Jupiter** rises around 8.30 pm, at magnitude −2.2 in Virgo.

Rising about 2 am, ringworld **Saturn** (magnitude +0.5) lies in Sagittarius.

Neptune lies too close to the Sun to be seen this month.

MOON

On **1 March**, the crescent Moon lies to the left of Venus, and immediately below Mars. The Moon moves in front of the Hyades

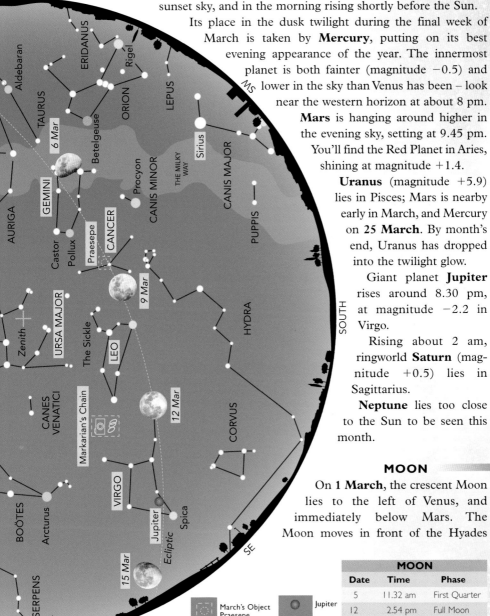

WEST

EAST

Aldebaran
ERIDANUS
TAURUS
Rigel
ORION
Betelgeuse
6 Mar
LEPUS
AURIGA
GEMINI
Castor
Pollux
Procyon
CANIS MINOR
Sirius
THE MILKY WAY
CANIS MAJOR
Praesepe
CANCER
PUPPIS
Zenith
URSA MAJOR
The Sickle
LEO
9 Mar
HYDRA
SOUTH
CANES VENATICI
Markarian's Chain
12 Mar
CORVUS
BOÖTES
Arcturus
VIRGO
Spica
15 Mar
Jupiter
Ecliptic
SE
SERPENS

March's Object
Praesepe

March's Picture
Markarian's Chain

Jupiter

Moon

MOON		
Date	Time	Phase
5	11.32 am	First Quarter
12	2.54 pm	Full Moon
20	3.58 pm	Last Quarter
28	3.57 am	New Moon

on **4 March**, occulting some of its brighter stars. The star near the Moon on **10 March** is Regulus. The Moon passes brilliant Jupiter on **14 March**, with Spica to the lower right. The Last Quarter Moon lies near Saturn on the morning of **20 March**. You'll find the slimmest crescent Moon to the left of Mercury on **29 March**. On **30 March**, the Moon lies to the left of Mars.

SPECIAL EVENTS

The Vernal Equinox, on **20 March** at 10.28 am, marks the beginning of spring, as the Sun moves up to shine over the northern hemisphere.

26 March, 1.00 am: British Summer Time starts – don't forget to put your clocks forward (the mnemonic is 'Spring forward, Fall back').

MARCH'S OBJECT

Between **Gemini** and **Leo** lies the faint zodiacal constellation of **Cancer** (the Crab). You'd be hard pressed to see it from a city, but concentrate your eyes on the crustacean's centre and, with a little luck, you can see a misty patch. This is **Praesepe** – a dense group of stars whose name literally means 'the manger', but is better known as the Beehive Cluster. If you train binoculars on it, you'll understand how it got its name: it really does look like a swarm of bees.

Praesepe lies nearly 600 light years away, and contains over 1000 stars, born together some 600 million years ago. Two of its stars have planets in orbit about them, but they are not 'Earths' – instead, they are 'hot Jupiters', gas giants circling close in to their parent star.

◀ *Eddie Guscott photographed Markarian's Chain – featuring (right) M86 and M84 – from Corringham, Essex. He captured this image using a 130 mm refractor, with a total exposure time of 5 hours 10 minutes through separate colour filters.*

◉ **OBSERVING TIP**
This is the ideal time of year to tie down your compass points – the directions of north, south, east and west as seen from your observing site. North is easy – just latch on to Polaris, the Pole Star (see this month's Constellation). And at noon, the Sun is always in the south. But the useful extra in March is that we hit the Spring (Vernal) Equinox, when the Sun rises due east, and sets due west. So remember those positions relative to a tree or house around your horizon.

Galileo, in 1610, was the first to recognize Praesepe as a star cluster. But the ancient Chinese astronomers obviously knew about it, naming the cluster *Zei She Ge* – 'the Exhalation of Piled-up Corpses'!

MARCH'S PICTURE

Markarian's Chain – named after Armenian astronomer Benik Markarian – is a line of galaxies in **Virgo** near the centre of the gigantic Virgo Cluster (see April's Constellation). Lying over 50 million light years away, the cluster is a swarm of 2000 galaxies. Though most are elliptical, in this image you'll spot three elegant, edge-on spiral galaxies. The Virgo Cluster is the hub of our local neck of the Universe, its mighty gravity even controlling the motion of our Milky Way.

MARCH'S TOPIC
Black holes

What happens to stars when they run out of their nuclear fuel and die? Lightweight specimens may thrive for trillions of years: yet massive stars – tens of times heavier than our Sun – are gas-guzzlers. They swallow up the contents of their stomachs in a mere few million years, collapsing internally, as their dwindling fuel can no longer sustain them. The result? A black hole.

A black hole is the densest concentration of matter in the Universe, and the most mysterious. For starters, light (and all other radiation) can't escape its gravity – so, by definition, it's black. And it's a hole. Nothing can escape its clutches: whatever falls in is trapped forever, because nothing can travel faster than light.

Black holes that result from a massive star's death are tiny; no more than 30 kilometres across. How do you find them in our mighty cosmos? Well … imagine a black cat having an altercation with a white cat in a dark coal cellar. You won't see the black cat. But from the way that the white cat is being swung around, you'll realize that there's a dark presence there: and you'll be able to hazard a guess as to how heavy it is.

And so it is with black holes. When one is in orbit around a normal star ('the white cat'), it drags streams of gas off its surface, and this material ends up in a superheated accretion disc circling the hole. The dazzling radiation from this disc is the giveaway.

So – what's the fate of the matter that falls into a black hole? Some scientists have calculated that it enters another universe: meaning that black holes could be gateways to a new cosmos.

Lording it over the April skies is giant planet Jupiter, at its closest to Earth this month. The brilliant world lies within the constellation **Virgo** (looking like a giant letter 'Y' in the heavens) who is cuddled up to her fellow spring constellation **Leo**, the celestial lion. And get out your binoculars to check out not one, but *two* comets!

▼ *The sky at 11 pm in mid-April, with Moon positions at three-day intervals either side of Full Moon. The star positions are also correct for midnight at the beginning of*

APRIL'S CONSTELLATION

The Y-shaped constellation of **Virgo** (the Virgin) is the second-largest in the sky. It takes a bit of imagination to see the group of stars as a virtuous maiden holding an ear of corn (the bright star **Spica**), but this very old constellation has associations with the time of harvest. In the early months of autumn, the Sun passes through the stars of Virgo, hence the connections with the gathering-in of fruit and wheat.

Spica is a hot, blue-white star over 12,000 times brighter than the Sun, boasting a temperature of 22,400°C. It has a stellar companion, which lies just 18 million kilometres away from Spica – closer than Mercury orbits the Sun. Both stars inflict a mighty gravitational toll on each other, raising enormous tides – creating two distorted, egg-shaped stars. In fact, Spica is the celestial equivalent of a rugby ball.

The glory of Virgo is in the 'bowl' of the Y-shape. Scan the upper region with a small telescope – at a low magnification – and you'll find it packed with faint, fuzzy blobs. These are just a few of the myriad galaxies – star-cities like the Milky Way – that make up the gigantic **Virgo Cluster** (see March's Picture), centred on the mammoth galaxy **M87**.

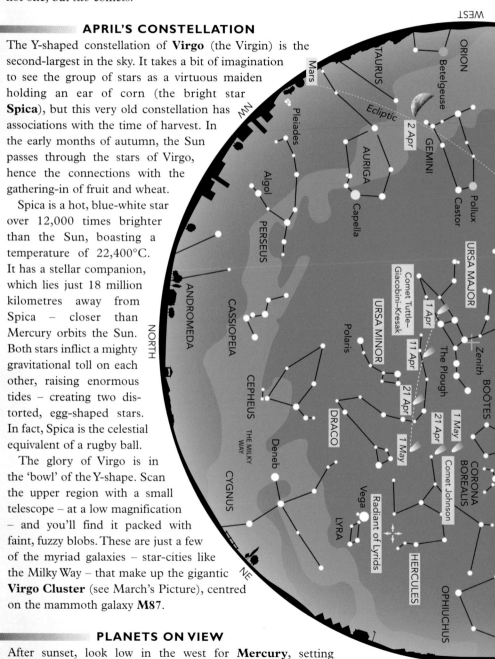

PLANETS ON VIEW

After sunset, look low in the west for **Mercury**, setting two hours after the Sun. The tiny world is at greatest eastern

April, and 10 pm at the end of the month. The planets move slightly relative to the stars during the month.

elongation on **1 April**, at magnitude 0, but fades rapidly as the month progresses.

Mars is on the move from Aries to Taurus, passing near the Pleiades on **21 April**. The Red Planet shines at magnitude +1.5 and sets around 11 pm.

The 'star' of the evening sky is **Jupiter**, which reaches opposition on **7 April** and is visible all night long. The giant world lies in Virgo, and shines far brighter than any star at magnitude −2.3. Through binoculars (held steadily) or a small telescope, watch the ever-changing antics of its four biggest moons.

Next up is **Saturn**, rising around 1 am in Sagittarius, at magnitude +0.4.

You may spot **Venus** (magnitude −4.3) in the morning twilight, rising an hour before the Sun.

Uranus and **Neptune** are lost in the Sun's glare this month.

MOON

The Moon lies near Regulus on **6 April**. On **10 April**, the Moon is close to Jupiter, with Spica below. You'll find the Moon to the left of Saturn on the morning of **17 April**. A thin crescent Moon hangs below Venus on the morning of **24 April**. Low in the dusk twilight on **28 April**, the crescent Moon appears near Aldebaran, with Mars to the right.

SPECIAL EVENTS

On **7 April**, Jupiter is at opposition.

Get your binoculars out for **Comet Tuttle–Giacobini–Kresak**! Though it

WEST

WEST

THE MILKY WAY

GEMINI

Procyon

CANIS MINOR

Castor
Pollux

CANCER

Alphard

HYDRA

5 Apr

The Sickle

URSA MAJOR

LEO

8 Apr

CORVUS

Zenith

CANES VENATICI

M87

Jupiter

Arcturus

Virgo Cluster

VIRGO

Spica

The Plough

CORONA BOREALIS

BOÖTES

SERPENS

11 Apr

Ecliptic

LIBRA

HERCULES

OPHIUCHUS

EAST

SE

SOUTH

MS

	April's Object and Picture	Mars
	Jupiter	Jupiter
	Radiant of Lyrids	Moon

MOON		
Date	**Time**	**Phase**
3	7.39 pm	First Quarter
11	7.08 am	Full Moon
19	10.57 am	Last Quarter
26	1.16 pm	New Moon

shines at only magnitude +7, the celestial wanderer is flying high in the sky, between the two bears – **Ursa Major** and **Ursa Minor** – and on towards the 'head' of **Draco** (the Dragon).

Towards the end of April, it's joined by **Comet Johnson** – another binocular object, brightening to magnitude +7 in **Hercules**.

21/22 April: It's the maximum of the **Lyrid** meteor shower, which appears to emanate from the constellation of Lyra. This is an excellent year for observing the Lyrids, as the Moon is well out of the way.

▲ *Damian Peach captured this image of Jupiter and its moons Io and Europa using a Celestron 14 telescope on 5 October 2013; the final photograph was compiled from 8 minutes of video sequences.*

APRIL'S OBJECT

At 143,000 kilometres in diameter, **Jupiter** is the biggest world in our Solar System. It could contain 1300 Earths – and the cloudy gas giant is very efficient at reflecting sunlight. Jupiter now is shining at a dazzling magnitude −2.3, and it's a fantastic target for stargazers, whether you're using your unaided eyes, binoculars, or a small telescope.

Despite its size, Jupiter spins faster than any other planet, in 9 hours 55 minutes. As a result, its equator bulges outwards

◉ **OBSERVING TIP**

It's always fun to search out the 'faint fuzzies' in the sky, and this month we have the whole **Virgo Cluster** of galaxies on display, as well as star clusters and nebulae sprinkled along the Milky Way in the northern sky. But don't even think about observing them near the time of Full Moon – its light will drown them out. You'll have the best views of these deep-sky objects around the time of New Moon – what astronomers call 'dark of Moon'. When the Moon is bright, though, there's still plenty to see: focus on planets, bright double stars – and, of course, the Moon itself. Plan your observing by checking the month-by-month Moon phases timetable in this book.

– through a small telescope, it looks like a squashed orange crossed with an old-fashioned humbug. The stripes are cloud belts of ammonia and methane stretched out by the planet's dizzy spin.

The planet's heart creates more energy than it receives from the Sun. The mysterious core of Jupiter is made of hydrogen, crushed down so it behaves like a liquid metal, and simmers at a temperature of 35,000°C. Had Jupiter been around 75 times more massive, its core would have been hot enough to fuse hydrogen to helium, in the process that powers the Sun: and the planet would have become a star.

Jupiter commands a family of almost 70 moons. The four biggest are visible in good binoculars, and even – to the really sharp-sighted – to the unaided eye.

Two new missions in the 2020s are planned to explore Jupiter's moons. The European Space Agency is building JUICE, a mission to probe three of Jupiter's icy moons: Ganymede, Europa and Callisto. And NASA will launch Europa, a spacecraft to make a reconnaissance of Jupiter's iciest moon, where scientists believe a giant under-surface ocean could lurk.

The driver for these missions is the search for water – because, if there's water, there could be primitive life.

APRIL'S PICTURE

Jupiter photographed with two of its major moons, volcanic Io and icy Europa, casting their shadows on the giant planet. The two moons could not be more different: violent Io with its active volcanoes; Europa with a possible ocean beneath its thick ice crust. Io (left) is markedly orange, even seen through a small telescope; Europa is pure white.

APRIL'S TOPIC
Date of Easter

The moveable feast of eggs and chocolate falls unusually late this spring, on **16 April**. OK – you may well ask why we're mentioning this religious event in our *Stargazing* book. In fact, the date of Easter has everything to do with astronomy: according to the Bible, Jesus was crucified at the Passover, whose date was fixed by the Jewish lunar calendar. So you can tell when it's Easter just by looking at the sky! First, wait until you see the Sun rising due east and setting due west: that's the Spring Equinox. Now follow the Moon until it's Full – and Easter will be the next Sunday. As a result, our chocolates may be brought out any time from 22 March to 25 April.

Of the stars on view in May, we have a soft spot for orange-coloured **Arcturus**. The brightest star in the constellation of **Boötes** (the Herdsman), it shepherds the two bears – **Ursa Major** and **Ursa Minor** – through the heavens, along with an accompanying dragon (**Draco**) and the superhero **Hercules**. Use binoculars to catch two faint comets tracking through this region of the sky.

▼ *The sky at 11 pm in mid-May, with Moon positions at three-day intervals either side of Full Moon. The star positions are also correct for midnight at the beginning of*

MAY'S CONSTELLATION

The triangular constellation of **Cepheus** is meant to represent the King of Ethiopia, married to the far more magnificent next-door constellation **Cassiopeia**. Both in legend and visually, his wife is far more exciting (she once boasted that her daughter **Andromeda** was more beautiful than all the sea nymphs, with almost disastrous effects). As a constellation, Cepheus is faint and somewhat boring – save for a trio of fascinating stars. **Alfirk** is a double star, with the companion being visible through a small telescope. The aptly known **Garnet Star** – named by William Herschel because of its ruddy hue – changes in brightness between magnitudes +3.4 and +5.1 with an approximate period of 2 years. But Cepheus is home to the most iconic of all variable stars – **Delta Cephei**. This star changes in brightness (from magnitude +3.5 to +4.4) over a period of 5 days and 9 hours. Astronomers discovered that this particular class of star (Cepheids) had variation timescales related to their intrinsic luminosities – allowing them to be used as pulsating stellar beacons to measure cosmic distances.

WEST

NW

Procyon
CANIS MINOR
HYDRA
Ecliptic
1 May
GEMINI
9 May
CANCER
Pollux
Castor
LEO
AURIGA
Capella
URSA MAJOR
NORTH
Algol
The Plough
Zenith
1 May
PERSEUS
Polaris
URSA MINOR
DRACO
1 May
CEPHEUS
11 May
Comet Tuttle-Giacobini-Kresak
CASSIOPEIA
Alfirk
Vega
ANDROMEDA
THE MILKY WAY
Delta Cephei
Garnet Star
Deneb
LYRA
AQUILA
CYGNUS
DELPHINUS
SAGITTA
Altair
NE
PEGASUS
EAST

PLANETS ON VIEW

Giant planet **Jupiter** is the undisputed master of the heavens, blazing all night long in Virgo at magnitude −2.2.

May, and 10 pm at the end of the month. The planets move slightly relative to the stars during the month.

At the beginning of the month, you can catch **Mars** (magnitude +1.6) low in the north-west after sunset, in Taurus, and setting at 10.45 pm. By the end of May, it's lost in the twilight glow.

Saturn rises around 11 pm, moving from Sagittarius into Ophiuchus in May. It shines at magnitude +0.2.

By the end of the month, faint **Neptune** (magnitude +7.9) reappears in the morning sky, rising around 2.30 am in Aquarius. Brilliant **Venus** lies low in the morning sky, at magnitude −4.3: it rises an hour-and-a-half before the Sun. **Uranus** and **Mercury** are too low in the dawn twilight to be seen this month, though the latter is at greatest western elongation on **17 May**.

MOON

The star near the Moon on **3 and 4 May** is Regulus. The Moon skims over Jupiter on **7 May** (see Special Events). On **8 May**, Spica lies to the lower right of the Moon, with Jupiter further to the right. The waning Moon is close to Saturn on the morning of **13 May**. Just before dawn on **22 May**, you'll find the crescent Moon to the right of Venus. The Moon rejoins Regulus on **31 May**.

SPECIAL EVENTS

Shooting stars from the Eta Aquarid meteor shower – tiny pieces shed by Halley's Comet burning up in Earth's atmosphere – fly across the sky on the night of **5/6 May**. Unfortunately, this year bright moonlight will wash out all but the brightest meteors.

WEST

CANCER

Regulus

The Sickle

4 May

HYDRA

URSA MAJOR

LEO

CORVUS

The Plough

CANES VENATICI

7 May

Jupiter

Zenith

Comet Johnson

BOÖTES

31 May

VIRGO

Spica

HYDRA

CENTAURUS

SOUTH

1 May

11 May

21 May

Arcturus

CORONA BOREALIS

SERPENS

LIBRA

HERCULES

OPHIUCHUS

10 May

SCORPIUS

AQUILA

THE MILKY WAY

Antares

SE

Altair

SERPENS

13 May

Saturn

EAST

Ecliptic

May's Object
Arcturus

Jupiter

Saturn

Moon

MOON		
Date	Time	Phase
3	3.47 pm	First Quarter
10	10.43 pm	Full Moon
19	1.33 am	Last Quarter
25	8.44 pm	New Moon

There's a gorgeous sight on **7 May**, when brilliant Jupiter lies immediately under the Moon.

And we have twin comets on binocular view in the northern sky, both starting May at magnitude +7.5, and only 15 degrees apart. **Comet Tuttle–Giacobini–Kresak** is moving from the head of **Draco** into **Hercules**, but rapidly fades to magnitude +9 by mid-month. **Comet Johnson**, on the other hand, gradually brightens to magnitude +6.5 as it tracks into **Boötes** by the end of May.

MAY'S OBJECT

As we said in the introduction, orange-red **Arcturus** in **Boötes** is one of our favourite stars in the sky. Its appearance is a sure sign that summer is on the way. The fourth brightest star in the heavens, Arcturus was a navigational beacon for Polynesian sailors, because it passed directly over Hawaii.

Arcturus is just entering its red giant phase, as it runs out of nuclear fuel at its core. Not much heavier than the Sun, it's more than 100 times as luminous. The star is slightly variable in brightness, which researchers put down to stellar oscillations: approaching the end of its life, Arcturus is distended and unstable.

◉ **OBSERVING TIP**

It's best to view your favourite objects when they're well clear of the horizon. If you observe them low down, you're looking through a large thickness of the atmosphere – which is always shifting and turbulent. It's like trying to observe the outside world from the bottom of a swimming pool! This turbulence makes the stars appear to twinkle. Low-down planets also twinkle – although to a lesser extent, because they subtend tiny discs, and aren't so affected.

Want to discover a comet? Look no further than your computer. Although most comets hurtle in from the outer Solar System, a few have turned out to be 'main belt' objects: comets lurking in the asteroid belt between Mars and Jupiter. Researchers want you to scrutinize images taken by Hawaii's huge Subaru Telescope (its mirror is 8.2 metres across). Your task will be to look at asteroids – which often turn up unexpectedly on images of the distant Universe – and check them out for cometary activity, like tails and outgassing from the nucleus. This will give astronomers a steer as to the connection between asteroids and comets.
https://www.zooniverse.org/projects/mschwamb/comet-hunters/

After it puffs off its inflated atmosphere, all that will remain of Arcturus will be its cooling, dying core – the nuclear reactor that once powered the star. Over time, this white dwarf will become a cold, black cinder.

MAY'S PICTURE

With two comets on view this month, now's the time to capture them on camera! These 'dirty snowballs' are relics from the earliest years of the Solar System: balls of ice and rock that flare up spectacularly as they plunge in towards the Sun. Being so ancient, they can reveal much about the origins of our local neighbourhood. And if a comet happens to pass close to us, we have the glorious view of a sensational sight in the sky: a dagger-shaped interloper trailing twin tails of gas and cosmic dust.

Early last year, our skies were graced by **Comet Catalina**. It never quite reached naked-eye brightness; but the comet was a grand sight in binoculars and telescopes, and – as you can see in this image – was a fantastic target for astrophotographers.

This year's comets won't be visible to the unaided eye, either, unless they surge unexpectedly in brightness. Comet Tuttle–Giacobini–Kresak was discovered by the three named astronomers on its appearances in 1858, 1907 and 1951. Comet Johnson was found in 2015 by a member of the Catalina Sky Survey.

◀ *Ian Sharp photographed Comet Catalina (C/2013 US10) from Rodalquilar, Spain, with a modified Canon 6D and Canon 200 mm L lens at f/4, ISO 3200, 7 × 3 minutes, on 17 January 2016. The galaxy M101 is at lower left and the well-known double star Mizar and Alcor at centre right.*

MAY'S TOPIC
The 13th constellation of the Zodiac

During the summer ahead, you'll find Saturn low in the south, in the constellation of **Ophiuchus**. But hang on – surely the Moon and planets only travel through the well-known constellations of the Zodiac, like Leo, Gemini or Taurus? Not entirely true. The ancient astronomers of Mesopotamia certainly noticed that the Sun, Moon and planets keep to a distinct band in the sky, and they divided it into distinctive star patterns. Because these 12 constella-tions largely depicted creatures – like a crab, a pair of fish and even a sea-goat – the Greeks called it the Zodiac, after their word for 'animal'. But if you look closely at a star map, you can see that this celestial highway only clips the top edge of **Scorpius** (the Scorpion), and actually runs through its more northerly neighbour, Ophiuchus (the Serpent Bearer). All the more reason to dismiss the mumbo-jumbo of the astrologers who claim the position of the Sun in the Zodiac at your birth determines your future: if you think you're a fierce scorpion, you're actually a guy carrying a snake!

It's low in the south, but grab a telescope if you can to view Saturn, putting on a fine display of its rings. Jupiter is brilliant in the evening sky; while Venus adorns the mornings this month. As the Sun reaches its highest position over the northern hemisphere, giving us the longest days and the shortest nights, June isn't the best month for seeing the fainter stars. But take advantage of the soft, warm weather to acquaint yourself with the lovely summer constellations of **Hercules**, **Scorpius**, **Lyra**, **Cygnus** and **Aquila**.

▼ The sky at 11 pm in mid-June, with Moon positions at three-day intervals either side of Full Moon. The star positions are also correct for midnight at the beginning of

JUNE'S CONSTELLATION

Down in the deep south of the sky this month lies a baleful red star. This is **Antares** – 'the rival of Mars' – and in its ruddiness it even surpasses the famed Red Planet. To ancient astronomers, Antares marked the heart of **Scorpius**, the celestial scorpion.

According to Greek myth, this summer constellation is intimately linked with the winter star pattern Orion, the great hunter who was killed by a mighty scorpion. The gods immortalized these two opponents as star patterns, placed at opposite ends of the sky so that Orion sets as Scorpius rises.

Scorpius is one of the few constellations that look like their namesakes. To the top right of Antares, a line of stars marks the scorpion's forelimbs. Originally, the stars we now call **Libra** (the Scales) were its claws. Below Antares, the scorpion's body stretches down into a fine curved tail (below the horizon on the chart), and deadly sting. Alas – the curved sting isn't visible from latitudes as far north as the UK. An excuse for a Mediterranean holiday!

Scorpius is a treasure-trove of astronomical goodies. Several lovely double stars include Antares: its faint companion looks greenish in contrast to Antares' strong red hue. Binoculars reveal the fuzzy patch of **M4**, a globular cluster

WEST

28 June

Regulus

Ecliptic

The Sickle

CANCER

LEO

Pollux

Castor

GEMINI

URSA MAJOR

CANES VENATICI

The Plough

AURIGA

URSA MINOR

DRACO

HERCULES

Zenith

Capella

Polaris

CEPHEUS

Vega

NORTH

CASSIOPEIA

LYRA

PERSEUS

THE MILKY WAY

CYGNUS

Algol

Deneb

ANDROMEDA

DELPHINUS

NE

Square of Pegasus

PEGASUS

EAST

June, and 10 pm at the end of the month. The planets move slightly relative to the stars during the month.

made of tens of thousands of stars, some 7,200 light years away.

The 'sting' contains two fine star clusters – **M6** and **M7** – so near to us that we can see them with the naked eye when they're well above the horizon: a telescope reveals their stars clearly.

PLANETS ON VIEW

The planetary 'star' of the month has to be **Saturn**, now at its closest and brightest – at magnitude +0.1. The ringworld reaches opposition on **15 June**, in Ophiuchus, and is visible all night long. Through a telescope Saturn forms a glorious sight, with its rings wider open this year than they have been since 2003 (see October's Special Events). On the night of **14/15 June**, the rings may appear unusually bright, as their icy particles reflect sunlight straight back at us (the 'opposition surge').

For sheer brilliance, though, Saturn is upstaged by **Jupiter**, at magnitude −2.0 in Virgo. The giant planet sets about 2 am.

The faintest planet, **Neptune** (magnitude +7.9), rises around 1 am in Aquarius. Towards the end of June, **Uranus** emerges in the morning sky, rising about 2 am. At magnitude +5.8, it lies in Pisces.

Venus is becoming more prominent in the pre-dawn sky. At magnitude −4.1, the Morning Star is rising just before 3 am. It reaches greatest western elongation on **3 June**.

Mercury and **Mars** are too close to the Sun to be visible this month.

MOON		
Date	Time	Phase
1	1.42 pm	First Quarter
9	2.09 pm	Full Moon
17	12.33 pm	Last Quarter
24	3.30 am	New Moon

Jupiter
Saturn
Moon

WEST

EAST

28 June
LEO
URSA MAJOR
CANES VENATICI
VIRGO
3 June
Jupiter
Spica
CORVUS
HYDRA
BOÖTES
Arcturus
Comet Johnson
1 July
Ecliptic
LIBRA
The Plough
Zenith
DRACO
CORONA BOREALIS
1 June
11 June
21 June
6 June
SERPENS
SCORPIUS
M4
HERCULES
OPHIUCHUS
Antares
Saturn
M6
Vega
LYRA
SAGITTA
SERPENS
9 June
SAGITTARIUS
M7
CYGNUS
Altair
AQUILA
THE MILKY WAY
Ecliptic
PEGASUS
DELPHINUS
AQUARIUS
CAPRICORNUS
SE
NW
SOUTH

MOON

You'll spot the Moon very close to Jupiter on **3 June**. On **4 June**, the Moon lies above Spica. The Full Moon passes over Saturn on the night of **9/10 June**. Venus and the crescent Moon make a fine pairing on the morning of **21 June**. The Moon is creeping up on Jupiter again on **30 June**.

SPECIAL EVENTS

Comet Johnson is putting on a good display through binoculars: view it after midnight, when the sky is really dark. The celestial wanderer starts the month at magnitude +6.5 in **Boötes**, gradually fading to magnitude +7 in **Virgo** by the end of June.

Venus reaches greatest western elongation on **3 June**.

On **15 June**, **Saturn** is at opposition.

21 June, 5.24 am: Summer Solstice. The Sun reaches its most northerly point in the sky, so 21 June is Midsummer's Day, with the longest period of daylight. Correspondingly, we have the shortest nights.

JUNE'S OBJECT

At the height of summer, the **Sun** rides high in the sky, and we feel the heat of its rays. Some 150 million kilometres away, it's our local star – and our local nuclear reactor.

At its core, where temperatures reach 15.7 million degrees, this giant ball of gas fuses hydrogen into helium. Every second,

CITIZEN SCIENCE
Sunspot mysteries
Help the scientists at Trinity College, Dublin, to understand the centuries-old enigma of sunspots – dark rashes on our local star where powerful magnetism breaks through. Despite the name, most are not single 'spots', but complex patterns of light and dark. Sunspotter provides thousands of close-ups from the SOHO solar space observatory, which you classify by their complexity. The goal is to learn how sunspots develop, and whether complex groups produce more eruptions – leading to more accurate forecasts of solar activity that can affect the Earth.
http://www.sunspotter.org/

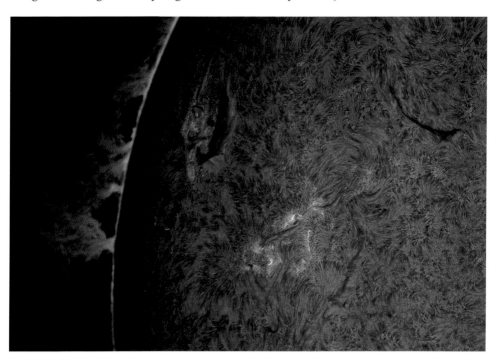

it devours 4 million tonnes of itself, bathing the Solar System with light and warmth.

Although it's now past the peak of its explosive activity, in a cycle that repeats roughly 11 years, the Sun is still a dangerous place. The driver is the Sun's magnetic field, wound up by the spinning of our star's surface gases. The magnetic activity suppresses the Sun's circulation, leading to a rash of dark sunspots. Then the pent-up energy is released in a frenzy, when our star hurls charged particles through the Solar System. If we are ever to make the three-year human journey to Mars, we will have to take the Sun's unpredictable, malevolent, weather into account.

The Sun's extreme heat makes it hazardous to observe without special precautions – follow the advice in this month's Observing Tip for a safe view of our local star.

JUNE'S PICTURE

When the **Sun** is very active, its powerful magnetism can suspend huge streamers of gas above its surface. You can see these sensational prominences yourself, using a solar telescope equipped with special filter (see Observing Tip). Prominences can erupt if their magnetic fields connect and short-circuit, in a titanic explosion of gases in the solar atmosphere. These coronal mass ejections, and other magnetic explosions on the Sun – including solar flares – can wreak havoc on Earth, causing power cuts and even knocking out stock markets!

JUNE'S TOPIC
Moon illusion

Ever wondered why the Moon appears bigger when it hangs low on the horizon? It's a pure illusion: nothing due to refraction through Earth's atmosphere. Although the atmosphere does affect the colour of the Moon (as it does in the case of the setting sun), it has no effect on the Moon's apparent size.

The effect has been known for millennia. Greek astronomers first noted it, and illustrious astronomers down the ages have written about the phenomenon. But it's not a physical effect; it's a trick played by the human eye. When you see the Moon low down, against a horizon of trees and houses, your mind compares it to these nearby objects – and so it appears larger. You can prove this by photographing the Moon at hourly intervals from rising to setting, then measuring the images. We guarantee that there'll be no change in size.

How to make the illusion go away? It's rumoured that standing with your back to the Moon, bending down, and looking at it between your legs, will make it appear normal again. Just don't let your friends catch you in the act ....

◀ *Pete Williamson captured this giant solar prominence on 23 April 2015 at 11:14 UT. He was using a Coronado 90 mm telescope equipped with an H-alpha filter.*

The brilliant trio of the **Summer Triangle** – the stars **Vega**, **Deneb** and **Altair** – is composed of the brightest stars in the constellations **Lyra**, **Cygnus** and **Aquila**. And this is the time to catch the far-southern constellations of **Sagittarius** and **Scorpius** – embedded in the glorious heart of the Milky Way.

▼ The sky at 11 pm in mid-July, with Moon positions at three-day intervals either side of Full Moon. The star positions are also correct for midnight at the beginning of

JULY'S CONSTELLATION

Lyra is small but perfectly formed. Shaped like a Greek lyre, it's dominated by brilliant white **Vega**, the fifth brightest star in the sky. Just 25 light years away – a near-neighbour in the cosmos – Vega is surrounded by a disc of dust that has probably given birth to baby planets.

Next to Vega is the **Double-Double**, a quadruple star known officially as epsilon Lyrae. Keen-sighted people can separate the pair, but you'll need a small telescope to find that each star is itself double.

The gem of Lyra lies between the two end stars of the constellation, **Sheliak** and **Sulafat**. The **Ring Nebula** needs a serious telescope (it's nearly ninth magnitude and little larger than Jupiter in apparent size), and is a wonderful example of a planetary nebula. Named by William Herschel, famed for his discovery of the planet Uranus, planetary nebulae look at first glance like dim, distant worlds. But in fact the Ring Nebula is a ghostly star corpse: the end of the road for a star like the Sun.

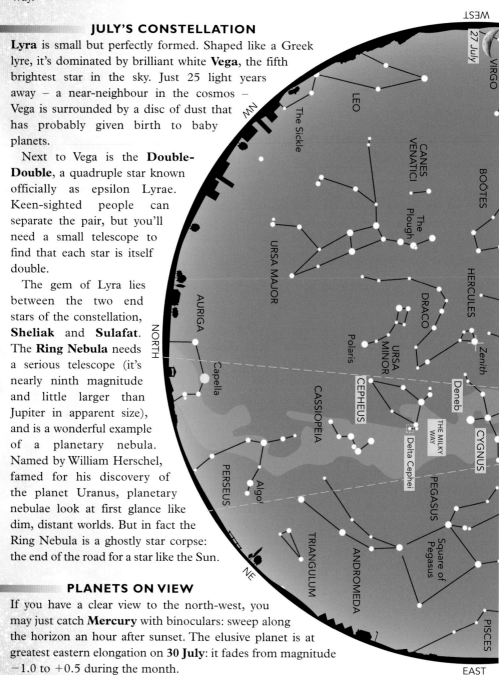

PLANETS ON VIEW

If you have a clear view to the north-west, you may just catch **Mercury** with binoculars: sweep along the horizon an hour after sunset. The elusive planet is at greatest eastern elongation on **30 July**: it fades from magnitude −1.0 to +0.5 during the month.

July, and 10 pm at the end of
the month. The planets move
slightly relative to the stars
during the month.

Jupiter, resplendent at magnitude −1.8, lies in Virgo and sets around midnight.

You'll find **Saturn** in Ophiuchus, at magnitude +0.2: the ringworld sets about 3 am. Use a small telescope to spot not only Saturn's rings, but also its largest moon, Titan.

Neptune (magnitude +7.8), in Aquarius, rises around 11 pm; while its near twin, **Uranus**, follows at midnight, shining at magnitude +5.8 in Pisces.

The Morning Star – **Venus** – is now unmistakeable, rising three hours before the Sun and shining at a brilliant magnitude −4.0. On the morning of **14 July**, it passes Aldebaran.

Mars is lost in the Sun's glare in July.

MOON

On **6 July**, the Moon lies near Saturn. You'll find the crescent Moon near Aldebaran on the morning of **20 July**, with Venus well to the left. On **25 July**, the crescent Moon is to the left of Mercury in the dusk twilight. The waxing Moon is close to Jupiter on **28 July**.

SPECIAL EVENTS

On **3 July**, at 9.11 pm, the Earth reaches aphelion, its furthest point from the Sun – 152 million kilometres out.

JULY'S OBJECT

At first glance, the star **Delta Cephei** – in the constellation representing King **Cepheus** – doesn't seem to merit any special attention. It's a yellowish star of mag-

WEST

27 July

Jupiter

Spica

VIRGO

3 July

LIBRA

MW

BOÖTES

Arcturus

CORONA BOREALIS

SERPENS

SERPENS

Antares

SCORPIUS

DRACO

Zenith

Vega

HERCULES

OPHIUCHUS

Saturn

6 July

SOUTH

Double-Double

LYRA

Sheliak

Sulafat

Ring Nebula

SERPENS

SAGITTA

THE MILKY WAY

SAGITTARIUS

Deneb

CYGNUS

SUMMER TRIANGLE

AQUILA

Altair

9 July

CAPRICORNUS

PEGASUS

DELPHINUS

Ecliptic

SE

PISCES

AQUARIUS

12 July

Neptune

EAST

Jupiter

Saturn

Neptune

Moon

[icon] July's Object
Delta Cephei

[icon] July's Picture
The Milky Way

MOON		
Date	**Time**	**Phase**
1	1.51 am	First Quarter
9	5.06 am	Full Moon
16	8.26 pm	Last Quarter
23	10.45 am	New Moon
30	4.23 pm	First Quarter

◄ *Robin Scagell used a fisheye lens to capture this view of the Milky Way arching over the skies of Guernsey in the Channel Islands. The Summer Triangle is to the left of centre, while the Plough and Polaris are at top right.*

CITIZEN SCIENCE
Map the Milky Way
– in 3D

Giant galaxies – like our Milky Way – are surrounded by a host of tinier, dwarf satellites. America's Rensselaer Polytechnic Institute started this project in 2009 specifically to investigate the Sagittarius Stream: the interaction between the Milky Way and the Sagittarius Dwarf Elliptical Galaxy. Now it's broadened out to map other star-streams. Join in, and help provide a broader picture of the 3D structure of our Galaxy, its formation and future evolution. These studies could even shed light on the nature of mysterious dark matter, which is crucial in the building and architecture of galaxies.
http://milkyway.cs.rpi.edu/milkyway/

nitude +4 – easily visible to the naked eye, but not prominent. A telescope reveals a companion star. But this star holds the key to measuring the size of the Universe.

Check the brightness of this star carefully over days and weeks, and you'll see that it changes regularly, from +3.5 (brightest) to +4.4 (faintest), every 5 days 9 hours. It's a result of the star literally swelling and shrinking in size, from 32 to 35 times the Sun's diameter.

⊙ *OBSERVING TIP*

This is the month when you really need a good, unobstructed horizon to the south, for the best views of the glorious summer constellations of Scorpius and Sagittarius. They never rise high in temperate latitudes, so make the best of a southerly view – especially over the sea – if you're away on holiday. A good southern horizon is also best for views of the planets, because they rise highest when they're in the south.

Astronomers have found that stars like this – Cepheid variables – show a link between their period of variation and their intrinsic luminosity. By observing the star's period and brightness as it appears in the sky, astronomers can work out a Cepheid's distance. With the Hubble Space Telescope, astronomers have now measured Cepheids in the galaxy NGC 4603, which lies 100 million light years away.

JULY'S PICTURE

The 'Via Lactea' – pathway of milk, or the **Milky Way** – arches across the summer skies. These are the more distant stars of our Galaxy, seen edge-on from our perspective. Its nature was unknown until Galileo swept the fuzzy band with his telescope. He described it as 'a congeries of stars' – a heap of distant suns. It's a glorious sight through binoculars: you can pick out individual stars, nebulae and star clusters. We'll be treated to great views of the Milky Way riding high in the sky from now until mid-winter.

JULY'S TOPIC
Centre of the Milky Way

The **Milky Way** stretches around our sky, as a gently glowing band. It's the inside view of a spiral-shaped galaxy of some 200–400 billion stars, with the Sun about halfway out. The centre of the Milky Way lies in the direction of Sagittarius, but the view – even for the most powerful optical telescopes – is blotted out by dense clouds of dark dust.

Now, telescopes observing infrared and radio waves have lifted the veil on the Galaxy's heart. They reveal stars and gas clouds whirling around at incredible speeds, up to 18 million km/h (12 million mph!), in the grip of something with fantastically strong gravity. It's cast-iron evidence for a supermassive black hole, weighing as much as 4 million Suns.

When a speeding star or gas cloud comes too close to this invisible monster at the Galaxy's heart, it's ripped apart. There's a final shriek – a burst of radiation – before it falls into the black hole, and disappears from our Universe.

Around 10 million years ago, the black hole feasted on a huge cloud of gas, and 'burped' two giant bubbles of superhot gas that have been detected by satellites observing gamma rays. More recently, it's been snacking on what may have been an asteroid that strayed too close in January 2015. The event generated a burst of X-rays 400 times more powerful than usual. Invisible to our eyes it may be, but the galactic centre is where the action is!

Whatever your holiday plans are – change them! Make sure you're in North America for the unforgettable sight of a total solar eclipse on **21 August**. Home or away, stay up late at night when the Moon's not around, to view the shimmering band of the Milky Way arching overhead.

▼ *The sky at 11 pm in mid-August, with Moon positions at three-day intervals either side of Full Moon. The star positions are also correct for midnight*

AUGUST'S CONSTELLATION

Capricornus (the Sea Goat) is one of a group of wet and watery constellations that swim in the celestial sea below **Pegasus** (the Winged Horse). But Capricornus had a special significance for ancient peoples. Over 2,500 years ago, the Sun nestled amongst its stars at the time of the Winter Solstice. It showed to them that the year was about to turn around: the hours of darkness were at an end, and life-giving spring was on the way. So this obscure triangle of faint stars may have been one of humanity's first constellations.

Algedi is the most interesting star in the constellation, lying at top right. Even with the unaided eye, you can see that the star is double. The pair are faint – magnitudes +3.6 and +4.6 respectively – and they aren't related. But each star is genuinely double, although you'll need a telescope to check this out.

By coincidence, the next-door star **Dabih** is also a double. The main member of this duo is a yellow star of magnitude +3.1; binoculars or a small telescope will reveal a blue companion at magnitude +6.

A telescope is also essential for the next beast in Capricornus – the globular cluster **M30**. The seventh-magnitude object, about 27,000 light years away, is just to the lower left of the constellation. This rather ragged ball of thousands of stars was probably among the first objects to form in our Galaxy. And it's very pretty – so, if you're into astrophotography, whether electronic or conventional – point and shoot!

at the beginning of August, and 10 pm at the end of the month. The planets move slightly relative to the stars during the month.

PLANETS ON VIEW

Jupiter shines in the west after sunset: at magnitude −1.7 in Virgo, the giant planet sets around 10 pm.

At magnitude +0.4, ringworld **Saturn** lies low in the south, in Ophiuchus, setting about 1 am.

Lying in Aquarius, **Neptune** (magnitude +7.8) rises around 9 pm, followed an hour later by **Uranus**, which lies in Pisces at magnitude +5.8.

Venus dominates the pre-dawn sky, rising about 2.30 am, at brilliant magnitude −3.9. Around **25 August**, you'll find the twin stars of Gemini – Castor and Pollux – directly above the Morning Star.

Mercury and **Mars** are too close to the Sun to be visible this month.

MOON

The Moon lies near Saturn on **2 and 3 August**. On the night of **15/16 August**, the Last Quarter Moon moves through the Hyades, occulting some of its brightest stars. A thin crescent Moon lies just below Venus on the morning of **19 August**. You'll find the waxing crescent Moon above Jupiter on **25 August**. The Moon revisits Saturn on **30 August**.

SPECIAL EVENTS

There's a partial eclipse of the Moon on **7 August**, visible from Africa and Asia. But by the time the Moon rises in the UK, it's just slipped out of the Earth's shadow. (More details at eclipse.gsfc.nasa.gov/lunar.html)

August's Object
Dumbbell Nebula

Radiant of Perseids

	MOON		
Date	**Time**	**Phase**	
7	7.10 pm	Full Moon	
15	2.15 am	Last Quarter	
21	7.30 pm	New Moon	
29	9.13 am	First Quarter	

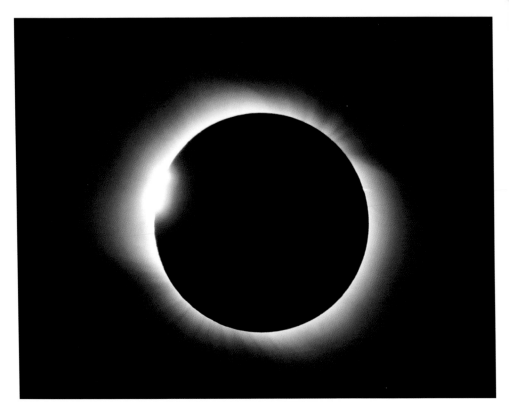

The maximum of the **Perseid** meteor shower falls on **12/13 August**, when the Earth runs into debris from Comet Swift–Tuttle. This year, the Moon will spoil the display after it rises at midnight.

It's going to be the most talked about astro-event of the year – and the most widely viewed online! On **21 August**, the track of a total solar eclipse sweeps right across the United States (see this month's Topic). A partial eclipse will be visible from all of North and Central America, and from northern regions of South America.

▲ *David Arditti photographed the total solar eclipse of 1 August 2008 from China. The corona surrounds the Sun's eclipsed disc, and – at left – the 'diamond ring' makes a beautiful appearance, with a portion of sunlight shining through a gap in the lunar mountains.*

AUGUST'S OBJECT

The tiny, faint constellation of **Vulpecula** (the Little Fox) contains one of the most beautiful sights in the sky – the **Dumbbell Nebula**. At magnitude +7.5, it is just visible through binoculars, and a real treat when seen through a telescope. Although an object of beauty, the Dumbbell is a denizen of doom. This 'planetary nebula' is the remains of a star that has died. The term was first coined by planet-discoverer William Herschel, who noted the resemblance of these circular nebulae to planets. It was his son John who nicknamed the Dumbbell, because it appeared to him 'like a double-headed shot…. A most amazing object.'

The Dumbbell lies roughly 1360 light years away (distances to planetary nebulae are notoriously difficult to measure), and is an estimated 2.5 light years across. It was created when an ageing star ran out of hydrogen fuel in its core, and puffed off its enveloping atmosphere into space. From the expansion rate of the nebula, astronomers estimate that the fatal act took place some 10,000 years ago.

Nevertheless, a corpse still lingers at the heart of the nebula. In its centre is a tiny white dwarf star: the collapsed core of the former star. It has no source of energy; all it can do is leak away its heat into space. Although it has a surface temperature of over 80,000°C at the moment, it will eventually end up as a cold, black cinder.

AUGUST'S PICTURE

We've chosen a stunning image of a past **solar eclipse** to give an idea of the treat in store for parts of North America this month (see Topic). Eclipses of the Sun are – amazingly – the result of sheer coincidence. The Sun is 400 times wider than the Moon; but it's also 400 times further away. When the paths of the two bodies cross, the Moon can overlap the Sun exactly.

AUGUST'S TOPIC
American solar eclipse

On **21 August 2017**, day will be turned to night across the mid-United States. For 90 minutes, the shadow of the Moon – blocking out the light from the Sun – will sweep across the US, from Oregon to South Carolina, at supersonic speeds. People near Nashville, Tennessee, will be treated to the longest duration of totality, at just over 2 minutes 40 seconds. It's the best eclipse in America since 1979.

As the Sun disappears, you'll feel increasingly damp and cold, while the birds stop singing. At the moment of totality, the sky goes black – apart from the Sun's corona, its pearly outer atmosphere, and perhaps some pink prominences. During the partial phases, you must use safe observing methods (see June's Observing Tip), but at totality it's briefly safe to view the eclipse with the naked eye or binoculars. Planets appear in the sky: look out for brilliant Venus well away to the west and Jupiter a similar distance to the east. Fainter Mars and Mercury are closer in; while Regulus lies right next to the eclipsed Sun.

Nothing prepares you for a total eclipse of the Sun. We've seen six, and we're still awed and terrified every time. So if you're in the United States in late August, don't miss out. (Details at eclipse.gsfc.nasa.gov/solar.html)

It's amazing to think that brilliant Jupiter – a feature of our evening skies all year – will be gone by the end of this month. And, as the nights grow longer than the days, the sky is now filling with the watery constellations of autumn: **Aquarius** (the Water Carrier), **Cetus** (the Sea Monster), **Capricornus** (the Sea Goat), **Pisces** (the Fishes), **Piscis Austrinus** (the Southern Fish) and **Delphinus** (the Dolphin).

▼ *The sky at 11 pm in mid-September, with Moon positions at three-day intervals either side of Full Moon. The star positions are also correct for midnight at*

SEPTEMBER'S CONSTELLATION

Although hardly one of the most spectacular constellations, **Aquarius** has a pedigree stretching back to antiquity, as it's slap-bang in the middle of a large group of 'watery' star patterns (described in the introduction above). The ancient Babylonians may have associated this zone of the heavens with water because the Sun passed through it during the rainy season, from February to March. The Greeks depicted Aquarius as a man pouring from a **Water Jar** (the central group of four faint stars), the liquid splashing downwards on to **Piscis Austrinus**.

Aquarius boasts one of the most glorious sights in the sky in long-exposure images; it's visible as a faint celestial ghost in binoculars or through a small telescope. Half the diameter of the Full Moon, the **Helix Nebula** is a star in its death throes. The Helix is a 'planetary nebula' – and, at 700 light years away, it's one of the nearest. Once a red giant star, the Helix is the result of the aged star puffing off its unstable, distended atmosphere into space – forming a beautiful spiral shroud around its collapsed core. This central white dwarf is a dead star bereft of nuclear power: it will gradually fade away to become a cold black cinder.

the beginning of September, and 10 pm at the end of the month. The planets move slightly relative to the stars during the month.

PLANETS ON VIEW

Brilliant **Jupiter** is slipping down into the twilight sky, setting only about an hour after the Sun. Lighting up Virgo, at magnitude −1.6, the giant planet has disappeared by the close of September.

Saturn, magnitude +0.5, lies among the dim stars of Ophiuchus, and sets around 10.30 pm.

Distant **Neptune** is at opposition on **5 September**. Visible all night long and at its nearest and brightest this year – though 'bright' means a measly magnitude +7.8 – you'll need binoculars or a telescope to spot the watery world in Aquarius.

Uranus (magnitude +5.7) rises at 8 pm in Pisces.

Rising around 3.30 am, **Venus** blazes at magnitude −3.8. For the first couple of nights of September, the Morning Star lies just below Praesepe, the star cluster in Cancer. On **20 September**, it's close to Regulus.

Just before dawn, **Mars** (magnitude +1.8) and **Mercury** (brightening from magnitude +1.2 to −0.8) are having a twilight tango in Leo, with its brightest star, Regulus. On **6 September**, Mars and Regulus are close, with Mercury to the right. Mars then slips downwards; Mercury passes close to Regulus on **10 and 11 September** and reaches greatest western elongation on **12 September**. On **16 and 17 September**, you'll find Mars and much brighter Mercury only 20 arc minutes apart.

WEST

SERPENS
OPHIUCHUS
HERCULES
SERPENS
Vega
LYRA
SAGITTA
Deneb
Altair
Zenith
CYGNUS
DELPHINUS
AQUILA
THE MILKY WAY
Ecliptic
SAGITTARIUS
28 Sept
3 Sept
CAPRICORNUS
CEPHEUS
Andromeda Galaxy
ANDROMEDA
PEGASUS
AQUARIUS
Square of Pegasus
Water Jar
Neptune
6 Sept
Helix Nebula
PISCIS AUSTRINUS
GRUS
SOUTH
Fomalhaut
TRIANGULUM
Uranus
PISCES
9 Sept
ARIES
Mira
CETUS
SE
TAURUS
ERIDANUS

EAST

Uranus
Neptune
September's Object
Andromeda Galaxy
Moon
September's Picture
The Moon

MOON		
Date	Time	Phase
6	8.03 am	Full Moon
13	7.25 am	Last Quarter
20	6.30 am	New Moon
28	3.54 am	First Quarter

◀ *Pete Lawrence, observing from Selsey in West Sussex, got up close and personal to the Moon in this shot. He took it at the prime focus of a C14 355 mm Schmidt-Cassegrain telescope using a Lumenera SKYnyx webcam-type camera.*

MOON

Look out for the crescent Moon with Venus and Regulus on the morning of **18 September** (see Special Events). Very low in the evening sky, the narrowest crescent Moon lies above Jupiter on **22 September**. On **26 September**, the planet near the Moon is Saturn.

SPECIAL EVENTS

Neptune reaches opposition on **5 September**.

On **15 September**, the Cassini space probe will crash into Saturn (see this month's Topic).

The crescent Moon forms a striking tableau as it hangs below brilliant Venus on the morning of **18 September**, with Regulus on the Moon's upper 'horn'.

The Autumn Equinox occurs at 9.02 pm on **22 September**. The Sun is over the Equator as it heads southwards in the sky, and day and night are equal.

SEPTEMBER'S OBJECT

Take the advantage of autumn's new-born darkness to pick out one of our neighbouring galaxies – the **Andromeda Galaxy**, sometimes known as M31 after its position in a catalogue of fuzzy patches compiled by Charles Messier. It covers an area four times bigger than the Full Moon. Like our Milky Way, it is a beautiful spiral shape, but – alas – it's presented to us almost edge-on.

The Andromeda Galaxy lies around 2.5 million light years away, and it's similar in size and shape to the Milky Way. It also

CITIZEN SCIENCE
Explore the Moon's secrets
NASA's Narrow-Angle Camera on its Lunar Reconnaissance Orbiter has provided the most detailed-ever images of the Moon's surface since it started circling our satellite in 2009. It can even see tyre-tracks made by the Apollo lunar rovers! Now the team wants your help. Moon Zoo asks you to investigate impact craters, to estimate the level of lunar bombardment; to identify unusual geological features, like lava channels and gullies; and to map the distribution of lunar boulders to help scientists decide upon safe future landing sites on the Moon. http://www.moonzoo.org/

hosts two bright companion galaxies – just as our own Galaxy does – as well as a flotilla of orbiting dwarf galaxies.

Unlike other galaxies, which are receding from us (as a result of the expansion of the Universe), the Milky Way and Andromeda are approaching each other. They will merge in about 5 billion years' time. The result of the collision may be a giant elliptical galaxy – nicknamed Milkomeda – which will be devoid of the gas that gives birth to new stars, and dominated by ancient red giants.

SEPTEMBER'S PICTURE

The Archimedes region of the **Moon**, photographed at First Quarter. Archimedes (left, 85 kilometres in diameter) is virtually in shadow. The two smaller craters to its east are Aristillus (top, 56 kilometres) and Autolycus (41 kilometres). The mountain range at the lower centre is known as the Lunar Apennines, with the broad lava plain of Mare Serenitatis (Sea of Serenity) to the right. The best time to observe lunar features is when the Sun highlights them from the side like this, enhancing their dramatic topography.

SEPTEMBER'S TOPIC
The great Cassini crash

NASA's space probe Cassini has been a constant companion of ringworld Saturn since 2004. Its first sensational success involved despatching Europe's Huygens spacecraft into the dense orange atmosphere of Saturn's biggest moon, Titan (visible through good binoculars), to land on its surface. Cassini has detected lakes of ethane and methane on Titan: possible breeding grounds for life when the Sun starts to enter its bulging red giant phase.

Since then, Cassini has investigated several of Saturn's moons (a total of 62 as of 2016), and the complex structure of its magnificent rings. The spacecraft also investigated a gigantic storm – the 'Great White Spot' of 2010 – an unusually active outburst on a generally inert planet.

Cassini's mission has been extended year after year, sending back invaluable data on the planet and its moons – not to mention glorious images. But all good things must come to an end. This year, the probe is being commanded to swing inside the rings, and to plummet to its destruction in Saturn's atmosphere on **15 September**.

A sad end for a plucky little spacecraft. But scientists are concerned that Cassini doesn't collide with – and pollute – any of Saturn's enigmatic moons: some of them could be the abodes of primitive life.

The glories of October's skies can best be described as 'subtle'. The barren square of **Pegasus** dominates the southern sky, with **Andromeda** attached to his side. But the dull autumn constellations are already being faced down by the brilliant lights of winter, spearheaded by the beautiful star cluster of the **Pleiades**. From Greece to Australia, these stars were seen as a gaggle of girls pursued by an aggressive male – **Aldebaran** or **Orion**.

▼ *The sky at 11 pm in mid-October, with Moon positions at three-day intervals either side of Full Moon. The star positions are also correct for midnight at*

OCTOBER'S CONSTELLATION

Pegasus is little more than a large, empty square of four medium-bright stars. But our ancestors somehow managed to see the shape of an upside-down winged horse here. In legend, Pegasus sprang from the blood of Medusa, the Gorgon, when **Perseus** (nearby in the sky) severed her head.

Scheat – the star at the top right of the **Square of Pegasus** – is a red giant over a hundred times wider than the Sun. Close to the end of its life, Scheat pulsates irregularly, changing in brightness by about a magnitude. Outside the square, **Enif** ('the nose') is a yellow supergiant. A small telescope, or even good binoculars, reveals a faint blue companion star.

Just next to Enif – and Pegasus's best-kept secret – is the beautiful globular cluster **M15**. You'll need a telescope for this one. M15 is around 33,000 light years away, and contains about 100,000 densely packed stars.

And Pegasus contains the first planet to be discovered beyond our Solar System, orbiting the star **51 Pegasi** (see this month's Topic).

PLANETS ON VIEW

Saturn is the only prominent planet in the evening sky; at magnitude +0.6 in Ophiuchus, it sets around 9 pm. This

WEST

OPHIUCHUS
AQUILA
CORONA BOREALIS
HERCULES
LYRA
THE MILKY WAY
CYGNUS
Vega
BOÖTES
DRACO
Deneb
CANES VENATICI
The Plough
CEPHEUS
Zenith
ANDROMEDA
NORTH
URSA MINOR
Polaris
CASSIOPEIA
URSA MAJOR
PERSEUS
Capella
AURIGA
TAURUS
Aldebaran
Castor
GEMINI
11 Oct
Ecliptic
Pollux
NE
Radiant of Orionids
Betelgeuse
ORION

EAST

the beginning of October, and 9 pm at the end of the month (after the end of BST). The planets move slightly relative to the stars during the month.

month, the famous rings appear wider open than they have in 14 years (see Special Events).

Setting at 3.30 am, faint **Neptune** (magnitude +7.8) lurks in Aquarius.

Uranus is at opposition on **19 October**, and visible all night long in Pisces. At magnitude +5.7, the seventh planet is just visible to the unaided eye – but binoculars are a big help.

In the morning sky, **Venus** is resplendent at magnitude −3.7, rising two hours before the Sun. On **5 and 6 October**, **Mars** lies just 20 arc minutes to the right of the Morning Star; but – at magnitude +1.8 – it's over a hundred times fainter. The Red Planet rises just before 5 am, and moves from Leo to Virgo during October.

Mercury and **Jupiter** are lost in the Sun's glare this month.

MOON

On **9 October**, the Moon lies near Aldebaran. The morning of **15 October** sees the crescent Moon next to Regulus. Before dawn on **17 October**, a narrow crescent Moon hangs well above Venus, with Mars in between. Look low in the dawn twilight of **18 October**, to spot the slimmest crescent Moon to the lower left of Venus. Back in the evening sky, you'll find the crescent Moon above Saturn on **24 October**.

SPECIAL EVENTS

In the early hours of **12 October**, a space-rock some 20 metres across will hurtle past the Earth, probably at less than one-tenth

WEST

SERPENS · THE MILKY WAY · AQUILA · 27 Oct · M_S · CAPRICORNUS · 2 Oct · PISCIS AUSTRINUS · SOUTH

CYGNUS · SAGITTA · DELPHINUS · Altair · AQUARIUS · Neptune · Fomalhaut

Deneb · M15 · Enif

CASSIOPEIA · Zenith · Scheat · 51 Pegasi · Square of Pegasus · PEGASUS · Ecliptic · 5 Oct · CETUS

ANDROMEDA · PISCES

PERSEUS · ARIES · Uranus · 8 Oct · Mira · ERIDANUS · SE

TRIANGULUM

Pleiades · Aldebaran · TAURUS · ORION · Rigel · Betelgeuse

EAST

October's Object
Uranus

Radiant of
Orionids

Uranus
Neptune
Moon

MOON		
Date	Time	Phase
5	7.40 pm	Full Moon
12	1.25 pm	Last Quarter
19	8.12 pm	New Moon
27	11.22 pm	First Quarter

43

the Moon's distance. Asteroid 2012 TC$_4$ should be visible in good binoculars for southern-hemisphere observers, whizzing past the Southern Cross.

Saturn's rings are at their maximum opening as viewed from Earth on **16 October**; we'll see them more and more obliquely until they appear edge-on in March 2025.

On **19 October**, Uranus is at opposition.

Debris from Halley's Comet smashes into Earth's atmosphere on **20/21 October**, causing the annual **Orionid** meteor shower. It's a great year for observing these shooting stars, as the Moon is well out of the way.

At 2 am on **29 October**, we see the end of British Summer Time for this year. Clocks go backwards by an hour.

OCTOBER'S OBJECT

If you're very sharp-sighted and have extremely dark skies, you stand a chance of spotting **Uranus** – the most distant planet visible to the unaided eye – at its closest to Earth this year on **19 October**. Discovered in 1781 by amateur astronomer William Herschel, Uranus was the first planet to be found since antiquity. The discovery doubled the size of our Solar System.

▼ *Damian Peach captured this image of Saturn and its rings on 16 December 2003, from Loudwater, Buckinghamshire. He used an 11-inch Schmidt-Cassegrain telescope working at f/31, coupled with an ATK-1 HS camera. Separate video sequences were recorded through red, green, blue and clear filters.*

Four times the diameter of the Earth, Uranus has an odd claim to fame: it orbits the Sun on its side (probably as a result of a collision in its infancy). Like the other gas giants, it has an encircling system of rings. But these are nothing like the spectacular edifices that girdle Saturn: Uranus's 13 rings are thin and faint. It also has a large family of moons: at the last count, 27.

The giant planet consists largely of a vast watery ocean surrounding a hot rocky core. The latest theories posit a hailstorm of diamonds dropping down through the water, to land in a sea of liquid diamond that wraps around the core. Though the Voyager 2 space probe revealed a bland, featureless world when it flew past Uranus in 1986, things are now hotting up as the planet's seasons change, with streaks and clouds appearing in its atmosphere.

OCTOBER'S PICTURE

Ringworld **Saturn**, with its glorious rings presented almost wide-open to us in 2003. Fourteen years on, the rings are once again at their most magnificent. Probably the remains of a moon torn up by Saturn's mighty gravity, Saturn's rings would stretch almost from the Earth to the Moon. Incredibly thin, the rings are made up of chunks of ice – ranging in size from ice cubes to refrigerators. They're an astonishing sight seen through a small telescope.

OCTOBER'S TOPIC
Extrasolar planets

The tally of planets circling other stars now stands at over 2000 – and thousands more wait to be confirmed.

The first came in 1995, when Swiss astronomers Michel Mayor and Didier Queloz discovered that the faint star **51 Pegasi** (see this month's chart) was being pulled backwards and forwards every four days. It had to be the work of a planet, tugging on its parent star. Astonishingly, this planet is around the same size as the Solar System's giant, Jupiter, but lies far closer to its star than Mercury. Astronomers call such planets 'hot jupiters'.

A team in California led by Geoff Marcy soon found more. Now astronomers are finding whole solar systems with several worlds in stable orbits – like the three (possibly four) orbiting the star Gliese 581. One of these planets is only a little more massive than the Earth, and may harbour liquid water.

The Kepler orbiting spacecraft, which picks up tiny diminutions in light when a planet crosses the disc of its parent star, has raked in new worlds, with nearly 5000 more candidates waiting to be checked out. The data garnered so far point to at least 17 billion Earth-sized planets in our Galaxy.

And now astronomers have captured their best image of an exoplanet. It circles the star Beta Pictoris, outside a disc of dust and debris that is forming into a new worlds. Researchers are waiting eagerly to see how the system evolves.

Take advantage of the moonless nights later this month to check out the most distant objects visible with the unaided eye. Away from streetlights, you should be able to see the misty blur of the great **Andromeda Galaxy**, the nearest large galaxy in the cosmos, lying 2.5 million light years away. And here's a challenge: spot the fainter **Triangulum Galaxy** – the light we see from this star-city left it almost 3 million years ago!

NOVEMBER'S CONSTELLATION

Perseus and its neighbour **Cassiopeia** are two of the best-loved constellations in the northern night sky. They never set as seen from Britain, and are packed with celestial goodies. In legend, Perseus was the superhero who slew Medusa, the Gorgon. Its brightest star, **Mirfak** ('elbow' in Arabic), lies 510 light years away and is 7000 times more luminous than our Sun.

But the 'star' of Perseus has to be **Algol** (whose name stems from the Arabic *al-Ghul* – meaning the 'demon star'). It represents the eye of Medusa – and it winks. Its variations were first reported by 18-year-old John Goodricke, a profoundly deaf amateur astronomer. He correctly surmised the 'winks' occur because a fainter star eclipses a brighter one (there are actually three stars in the system). Over a period of 2.9 days, Algol dims from magnitude +2.1 to +3.4.

Another gem in Perseus – or to be exact, two of them – is the **Double Cluster**, h and chi Persei. Visible to the unaided eye, the duo is a sensational sight in binoculars. Some 7500 light years distant, the clusters are made of bright young blue stars. Both are a mere 12.8 million years old (compare this to our Sun, which has notched up 4.6 *billion* years so far!).

▼ The sky at 10 pm in mid-November, with Moon positions at three-day intervals either side of Full Moon. The star positions are also correct for 11 pm at

PLANETS ON VIEW

the beginning of November, and 9 pm at the end of the month. The planets move slightly relative to the stars during the month.

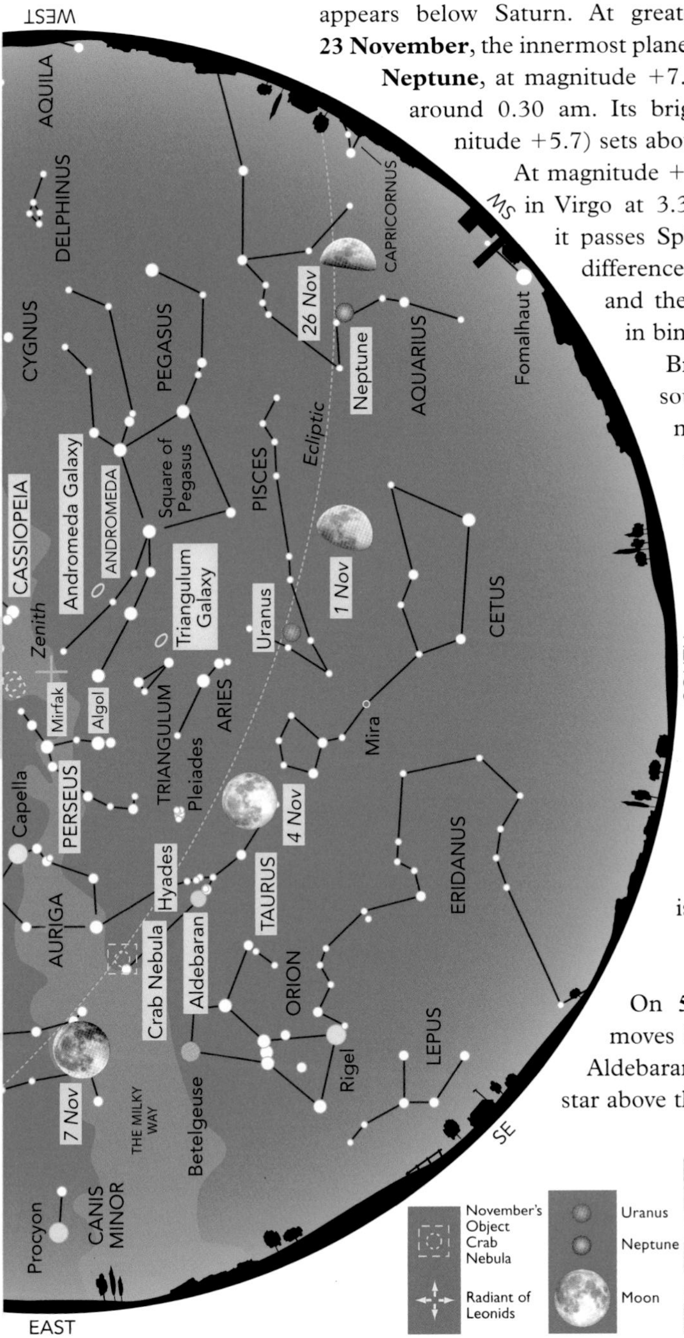

Stately **Saturn** is descending into the evening twilight, and sets around 6 pm. At magnitude +0.6, the ringworld is spectacular through a small telescope as it moves from Ophiuchus into Sagittarius.

During the last few days of November, brighter **Mercury** appears below Saturn. At greatest eastern elongation on **23 November**, the innermost planet shines at magnitude −0.2. **Neptune**, at magnitude +7.9 lies in Aquarius and sets around 0.30 am. Its brighter twin, **Uranus** (magnitude +5.7) sets about 4.30 am in Pisces.

At magnitude +1.7, you'll find **Mars** rising in Virgo at 3.30 am. On **30 November**, it passes Spica – check out the colour difference between the Red Planet and the blue-white star (best seen in binoculars).

Brilliant **Venus** appears in the south-east around 6 am, at magnitude −3.8. The star to the lower right on **2 and 3 November** is Spica. As Venus slips down into the dawn glow, it passes very close to Jupiter on **13 November** (see Special Events).

Jupiter is moving upwards in the morning twilight, shining at magnitude −1.5 on the borders of Virgo and Libra. By the end of November, the giant planet is rising as early as 5 am.

MOON

On **5/6 November**, the Moon moves in front of the Hyades and Aldebaran (see Special Events). The star above the Moon on the morning of

Star chart labels

WEST · AQUILA · DELPHINUS · CYGNUS · PEGASUS · CAPRICORNUS · 26 Nov · Neptune · AQUARIUS · Fomalhaut · Ecliptic · Square of Pegasus · PISCES · CASSIOPEIA · Andromeda Galaxy · ANDROMEDA · Zenith · Triangulum Galaxy · Uranus · 1 Nov · CETUS · Mirfak · Algol · TRIANGULUM · ARIES · Mira · Capella · PERSEUS · Pleiades · 4 Nov · ERIDANUS · AURIGA · Hyades · TAURUS · SOUTH · Crab Nebula · Aldebaran · ORION · LEPUS · THE MILKY WAY · Betelgeuse · Rigel · 7 Nov · CANIS MINOR · Procyon · SE · EAST

Legend

- November's Object Crab Nebula
- Radiant of Leonids
- Uranus
- Neptune
- Moon

MOON		
Date	Time	Phase
4	5.23 am	Full Moon
10	8.36 pm	Last Quarter
18	11.42 am	New Moon
26	5.03 pm	First Quarter

12 **November** is Regulus. On the morning of **15 November**, the crescent Moon is next to Mars. The slimmest crescent Moon lies to the left of Venus and Jupiter low in the dawn glow on **17 November** (best seen in binoculars). Back in the evening twilight, you may catch the crescent Moon next to Saturn on **20 November**, with Mercury below.

SPECIAL EVENTS

In between the fireworks on **5/6 November**, turn your eyes (and preferably binoculars) skywards to watch a fine occultation of the **Hyades** by the Moon, starting just after moonrise and ending when the Moon occults **Aldebaran** around 2.30 am (exact time depending on your location).

Very low in the dawn twilight on **13 November**, the two brightest planets – Venus and Jupiter – lie only 15 arc minutes apart: a gorgeous sight in binoculars or a small telescope.

Keep an eye open on **16/17 November** for possible brilliant shooting stars from debris that Comet Tempel–Tuttle off-loaded as it passed the Sun in 1300. We are treated to the regular peak of the **Leonid** meteor shower the following night, **17/18 November**, when Earth runs into the main stream of fragments from this comet. It promises to be a good year, as moonlight won't interfere.

NOVEMBER'S OBJECT

This month, we home in on a small region above the 'lower horn' of **Taurus** (the Bull). Here, in 1054, Chinese astronomers witnessed the appearance of a brilliant 'guest star'. Visible in daylight for 23 days, the supernova remained in the night sky for nearly two years. But it was no stellar debutante

CITIZEN SCIENCE
Find your own pulsar
Imagine you could discover gravitational waves – last year's big astronomy news story – at home! Einstein@Home provides your computer with data from the LIGO gravitational wave detectors, that it can number-crunch in its idle moments. The project is on the track of spinning neutron stars, called pulsars: as well as the gravitational-wave survey, you're provided with observations from the Fermi gamma-ray satellite and from the world's biggest radio telescope, in Arecibo, Puerto Rico. Already, home volunteers have discovered dozens of previously unknown pulsars.
https://einstein.phys.uwm.edu/

⊙ **OBSERVING TIP**

The Andromeda Galaxy is often described as the furthest object 'easily visible to the unaided eye'. But it can be a bit elusive – especially if you are suffering from light pollution. The trick is to memorize Andromeda's pattern of stars, and then to look slightly to the *side* of where you expect the galaxy to be. This technique – called 'averted vision' – causes the image to fall on the outer region of your retina, which is more sensitive to light than the central region that's evolved to discern fine details. You'll certainly need averted vision to track down Andromeda's fainter sibling, the Triangulum Galaxy, with the naked eye. The technique is also crucial when you want to observe the faintest nebulae or galaxies through a telescope.

◀ *Robin Scagell caught this aurora from Svolvaer, in the Lofoten Islands, Norway, on 12 March 2011. Photographed from MS Marco Polo, Robin used a 17 mm lens to make an 8-second exposure at ISO 1600.*

– it was an old star on the way out, exploding because it was overweight.

Today, we see the remnants of this supernova as the **Crab Nebula** – named by 19th-century Irish astronomer the Earl of Rosse because it resembled a crab's pincers. The expanding debris now measures 11 light years across. You can just make out the Crab Nebula through a small telescope, but it is small and faint (magnitude +8.4).

At the centre of the Crab Nebula is the core of the dead star, which has collapsed to become a pulsar. This tiny, but super-dense object – only the size of a city, but with the mass of the Sun – is spinning around furiously at 30 times a second and emitting beams of radiation like a lighthouse.

NOVEMBER'S PICTURE

It's never been so trendy to go on a cruise to view the **aurora borealis**. The 'merrie dancers', as the Scots call them, are the result of solar activity. Although our local star is calming down now, after its recent bout of magnetic agitation – sunspots, flares and coronal mass ejections – it continues to throw electrically charged particles towards our planet that get channelled towards our north and south magnetic poles. The result? A glorious light show: these particles hit the Earth's atmosphere, and light up the skies like gas in a neon tube. The heavens become alive with shifting red and green curtains. Not predictable – but, if you get the chance (check the web), they're not to be missed!

NOVEMBER'S TOPIC
Constellations

Perseus, our constellation of the month, highlights our obsession to 'join up the dots' in the sky and weave stories around them. One explanation is that the patterns acted as an *aide-mémoire* to ancient farming communities, as the seasons altered in step with annual change in the constellations as the Earth moved around the Sun.

The stars were also a great steer to navigation at sea. Greek astronomers may have 'mapped' their legends on to the sky, so that sailors crossing the Mediterranean would associate key constellations with their traditional stories.

Not all the world saw the heavens through western eyes. The Chinese divided up the sky into a plethora of tiny constellations, containing three or four stars apiece. And the Australian Aborigines, in their dark deserts, were so overwhelmed with stars that they made constellations out of dark patches where they couldn't see any!

Brave the winter chill to enjoy the celestial pyrotechnics of the Geminid meteor shower, along with some scintillating constellations. **Orion**, with his hunting dogs **Canis Major** and **Canis Minor**, is fighting his adversary **Taurus** (the Bull). The all-powerful hunter is accompanied a pair of hero twins in **Gemini**, and the charioteer **Auriga** almost overhead. Look more closely for faint **Lepus**, the timid hare cowering below Orion's feet.

▼ The sky at 10 pm in mid-December, with Moon positions at three-day intervals either side of Full Moon. The star positions are also correct for 11 pm at

DECEMBER'S CONSTELLATION

Sparkling overhead, **Auriga** (the Charioteer) is named after the lame Greek hero Erichthoneus, who invented the four-horse chariot. The constellation's roots date way back to the ancient Babylonians, who saw Auriga as a shepherd's crook.

Capella, the sixth brightest star, means 'the little she-goat'; but there's nothing modest about this giant yellow star, which is 150 times more luminous than our Sun, twice as wide, and almost three times heavier. It also holds a substantial yellow companion in thrall. More controversially, it may be orbited by two faint red dwarf stars.

Nearby, you'll find a tiny triangle of stars nicknamed 'the Kids' (Haedi). Two are eclipsing binaries: stars that change in brightness because a companion passes in front. **Zeta Aurigae** is an orange star eclipsed every 972 days by a blue partner.

Epsilon Aurigae is a weirdo. Every 27 years, it suffers two-year-long eclipses, caused by a dark disc of material as big as the orbit of Jupiter. No two eclipses are the same: and there are tantalizing hints of giant proto-planets within the disc.

Also, bring out those binoculars (better still, a small telescope) to sweep within the 'body' of the Chariot to find three very pretty open star clusters: **M36**, **M37** and **M38**.

EAST

the beginning of December, and 9 pm at the end of the month. The planets move slightly relative to the stars during the month.

PLANETS ON VIEW

Only two dim planets are visible in the evening sky: **Neptune** (magnitude +7.9) in Aquarius, setting about 10.30 pm; and **Uranus**, which sets around 2.30 am and lies in Pisces at magnitude +5.7.

But the planetary pace hots up after midnight. First on the scene – at 3.30 am – is **Mars**, at magnitude +1.6 and moving from Virgo into Libra.

It's followed by **Jupiter**, which rises at 4.30 am mid-month. The giant planet shines at magnitude −1.6 in Libra; it's gradually moving up in the morning sky, and approaches Mars by the year's end.

Right at the start of December, you may catch **Venus** (magnitude −3.8) very low in the dawn sky – rising an hour before the Sun – but it soon drops down into the twilight zone.

In the final few days of 2017, its place is taken by **Mercury**: the innermost planet is visible just before 7 am, low in the south-east and shining at magnitude −0.1.

Saturn is lost in daylight this month.

MOON

The star next to the Full Moon on **3 December** is Aldebaran. On **8 December**, the rising Moon occults Regulus (see Special Events). On the morning of **13 December**, the crescent Moon forms a triangle with Mars and Spica, with Jupiter well below. You'll find a slim crescent Moon near brilliant Jupiter on the mornings of **14 and 15 December**; Mars lies to the upper right.

MOON		
Date	Time	Phase
3	3.47 pm	Full Moon
10	7.51 am	Last Quarter
18	6.30 am	New Moon
26	9.20 am	First Quarter

On the night of **30/31 December**, the Moon occults the Hyades and Aldebaran (see Special Events).

SPECIAL EVENTS

The Full Moon of **3 December** is the biggest and brightest of the year – but it's upstaged by next month's 'supermoon,' on **2 January 2018**.

As the Moon rises on **8 December**, it lies right in front of Regulus. The occultation ends at around 10.15 pm, when you'll see the bright star pop into view at the Moon's dark limb (the exact time depends on your location).

The Moon is well out of the way, so enjoy the year's most prolific and spectacular meteor shower, the **Geminids**, on **13/14 December** (see this month's Object).

The Winter Solstice occurs at 4.28 pm on **21 December**. As a result of the tilt of Earth's axis, the Sun reaches its lowest point in the heavens as seen from the northern hemisphere: we get the shortest days, and the longest nights.

As it grows dark on **30 December**, you'll find the Moon in front of the **Hyades**. Around 1 am on **31 December** (the exact time depends on your location), the Moon occults **Aldebaran**, the star reappearing just under an hour later.

DECEMBER'S OBJECT

We're due for serious celestial fireworks on the night of **13/14 December**: the **Geminid** meteor shower, spitting out shooting stars from the direction of the Heavenly Twins, **Gemini**. From a dark location, you may spot 100 meteors an hour. They're best seen after midnight, when the Earth comes face-on to the meteor stream. Wrap up warmly! And just take your eyes with you: you don't need a telescope or binoculars.

Most meteors are dusty debris from comets, burn-

▼ *Robin Scagell captured this Geminid meteor on 14 December 1996, using a 24 mm lens at f/2.8, with ISO 1600 film and a 3-minute exposure. He was observing from Tillingham, Essex. The light pollution comes from the north Kent coast, some 60 km away.*

⊙ **OBSERVING TIP**

Hold a 'meteor party' to check out the year's best celestial firework show, the Geminid meteor shower on 13/14 December. You don't need any optical equipment – in fact, telescopes and binoculars will restrict your view of the shooting stars, which can appear anywhere. The ideal viewing equipment is your unaided eye, plus a warm sleeping bag and a lounger. Everyone should look in different directions, so you can cover the whole sky: shout out 'Meteor!' when you see a shooting star. One of the party can record the observations, using a watch, notepad and red torch. In the interests of science, try to brave the cold and observe the sky for at least an hour, before repairing indoors for some warming seasonal cheer.

ing up in Earth's atmosphere at high speeds. But the 'parent body' of the Geminids was for long a mystery. Then, in 1983, astronomers using the pioneering Infrared Astronomical Satellite (IRAS) – designed to track down heat radiation from distant newborn stars – picked up a fast-moving asteroid in the Solar System.

Its orbit was wild, bringing it closer to the Sun than Mercury. There was only one possible name for this wayward celestial beast: Phaethon – the mythological lad who disastrously tried to drive the sun-chariot across the sky.

The IRAS team realized that the real Phaethon's orbit matches the path of the Geminid meteors. This wayward asteroid is most likely a 'dead comet', whose original ices have boiled away after repeated close encounters with the Sun. But the odd flare-up suggests there may be life in the old dirty snowball yet....

Being more substantial than a fresh comet, Phaethon yields chunkier meteors. So look out for some beautiful, brilliant, slow-moving shooting stars this month.

DECEMBER'S PICTURE

A **Geminid meteor** shooting past Orion. Backtrack its path, and you'll see that it comes from the region of Gemini, to the top left of Orion.

DECEMBER'S TOPIC
Stonehenge: midwinter monument

Today's Druids make it easy for themselves: they hold their Stonehenge ceremonies in the warmth of June. But its builders were actually celebrating chilly Midwinter's Day. Instead of standing inside Stonehenge, watching the summer Sun rising over the outlying Heel Stone, they were stationed at the Heel Stone, observing the pale winter Sun setting through the stone arches.

Archaeologists have found natural grooves in the bedrock near the Heel Stone, that point towards Stonehenge and would have naturally guided our ancestors' eyes in that direction. And the clinching evidence comes from the remains of great annual feasts held nearby, where thousands of people converged from all over the country. These include the bones of pigs that were killed and eaten there when they were nine months old. Piglets are naturally born in the spring, so the great celebrations at Stonehenge must have been timed for Midwinter.

So — forget the Druids: be at Stonehenge for sunset on **21 December**!

There's always something to see in our Solar System, from planets to meteors or the Moon. These objects are very close to us – in astronomical terms – so their positions, shapes and sizes appear to change constantly. It is important to know when, where and how to look if you are to enjoy exploring Earth's neighbourhood. Here we give the best dates in 2017 for observing the planets and meteors (weather permitting!), and explain some of the concepts that will help you to get the most out of your observing.

THE INFERIOR PLANETS

A planet with an orbit that lies closer to the Sun than the orbit of Earth is known as *inferior*. Mercury and Venus are the inferior planets. They show a full range of phases (like the Moon) from the thinnest crescents to full, depending on their position in relation to the Earth and the Sun. The diagram below shows the various positions of the inferior planets. They are invisible when at *conjunction*, when they are either behind the Sun, or between the Earth and the Sun, and lost in the latter's glare.

Magnitudes

Astronomers measure the brightness of stars, planets and other celestial objects using a scale of *magnitudes*. Somewhat confusingly, fainter objects have higher magnitudes, while brighter objects have lower magnitudes; the most brilliant stars have negative magnitudes! Naked-eye stars range from magnitude −1.5 for the brightest star, Sirius, to +6.5 for the faintest stars you can see on a really dark night.

As a guide, here are the magnitudes of selected objects:

Sun	−26.7
Full Moon	−12.5
Venus (at its brightest)	−4.7
Sirius	−1.5
Betelgeuse	+0.4
Polaris (Pole Star)	+2.0
Faintest star visible to the naked eye	+6.5
Faintest star visible to the Hubble Space Telescope	+31

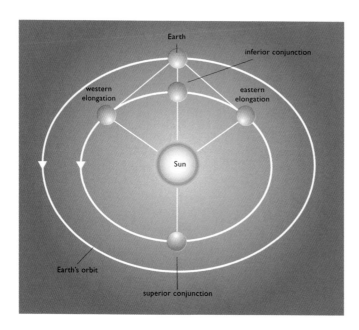

◀ At eastern or western elongation, an inferior planet is at its maximum angular distance from the Sun. Conjunction occurs at two stages in the planet's orbit. Under certain circumstances, an inferior planet can transit across the Sun's disc at inferior conjunction.

Mercury

In the evening sky, we'll have our best views of this elusive planet at the end of March and the start of April; Mercury's evening appearances in July and November are low in bright twilight skies. The innermost planet puts on good morning shows in January, September and December (though it's lost in the dawn twilight at the May apparition).

⬤ Maximum elongations of Mercury in 2017	
Date	Separation
19 January	24° west
1 April	19° east
17 May	26° west
30 July	27° east
12 September	18° west
23 November	22° east

Maximum elongation of Venus in 2017	
Date	Separation
12 January	47° east
3 June	46° west

Venus
From January to mid-March, Venus is a glorious Evening Star. On **25 March**, it swings between Sun and Earth – for a few days visible in both the evening and morning sky – before taking up residence as the Morning Star for the rest of 2017.

THE SUPERIOR PLANETS
The superior planets are those with orbits that lie beyond that of the Earth. They are Mars, Jupiter, Saturn, Uranus and Neptune. The best time to observe a superior planet is when the Earth lies between it and the Sun. At this point in a planet's orbit, it is said to be at *opposition*.

▶ *Superior planets are invisible at conjunction. At quadrature the planet is at right angles to the Sun as viewed from Earth. Opposition is the best time to observe a superior planet.*

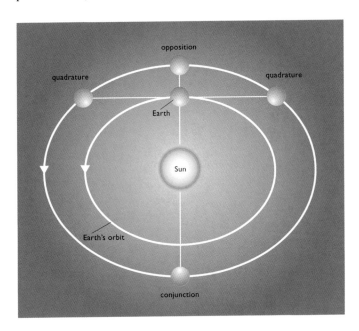

● Progress of Mars through the constellations	
Early January	Aquarius
Mid-Jan to February	Pisces
March to mid-April	Aries
Mid-April to May	Taurus
September to mid-Oct	Leo
Mid-Oct to mid-Dec	Virgo
Late December	Libra

Mars
The Red Planet doesn't reach opposition in 2017, but it's riding high in the evening sky until May; and is visible in the morning sky from September until the end of the year.

Jupiter
The giant planet blazes in the evening sky at the start of 2017. It inhabits Virgo from January, through its opposition on **7 April**, until it disappears into the evening twilight glow at the end of September. Jupiter reappears in the morning sky in November, moving from Virgo to Libra.

Saturn
The ringed planet spends the year low in the sky, on the border of Ophiuchus and Sagittarius. Saturn is prominent in

the morning sky at the start of 2017 and is visible all night when it reaches opposition on **15 June**. On **16 September** the famous rings appear wider open than they have done since 2003. Saturn slips into the evening twilight towards the end of November.

Uranus

Just perceptible to the naked eye, Uranus is visible in the evening sky from January to March; and then from June to December. It swims among the stars of Pisces all year, and is at opposition on **19 October**.

Neptune

Lying in Aquarius all year, the most distant planet is at opposition on **5 September**. Neptune can be seen – though only through binoculars or a telescope – in January; and then from May to the end of the year.

SOLAR AND LUNAR ECLIPSES

Solar Eclipses

The track of an annular eclipse on **26 February** passes from southern Chile and Argentina, across the South Atlantic Ocean, to Angola. People in southern regions of South America and western Africa will experience a partial solar eclipse.

It's the big one! On **21 August**, the track of a total solar eclipse sweeps across the United States, from Oregon to South Carolina. A partial eclipse will be visible in all of northern and Central America, and from northern regions of South America.

Lunar Eclipses

A partial eclipse of the Moon on **7 August** is visible from Africa and Asia. But by the time the Moon rises in the UK, it's just slipped out of the Earth's shadow.

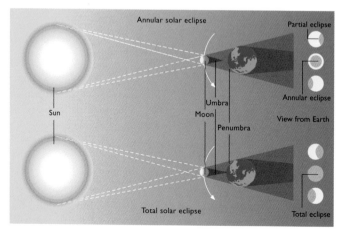

◀ *Where the dark central part (the umbra) of the Moon's shadow reaches the Earth, we see a total eclipse. People located within the penumbra see a partial eclipse. If the umbral shadow does not reach the Earth, we see an annular eclipse. This type of eclipse occurs when the Moon is at a distant point in its orbit and is not quite large enough to cover the whole of the Sun's disc.*

Dates of maximum for selected meteor showers	
Meteor shower	Date of maximum
Quadrantids	3/4 January
Lyrids	21/22 April
Eta Aquarids	5/6 May
Perseids	12/13 August
Orionids	20/21 October
Leonids	16/17 November, 17/18 November
Geminids	13/14 December

▶ *Meteors from a common source, occurring during a shower, enter the atmosphere along parallel trajectories. As a result of perspective, however, they appear to diverge from a single point in the sky – the radiant.*

Angular separations

Astronomers measure the distance between objects, as we see them in the sky, by the angle between the objects in degrees (symbol °). From the horizon to the point above your head is 90 degrees. All around the horizon is 360 degrees.

You can use your hand, held at arm's length, as a rough guide to angular distances, as follows:

Width of index finger   1°
Width of clenched hand 10°
Thumb to little finger
   on outspread hand   20°
For smaller distances, astronomers divide the degree into 60 arc minutes (symbol ′), and the arc minute into 60 arc seconds (symbol ″).

METEOR SHOWERS

Shooting stars – or *meteors* – are tiny particles of interplanetary dust, known as *meteoroids*, burning up in the Earth's atmosphere. At certain times of year, the Earth passes through a stream of these meteoroids (usually debris left behind by a comet) and we see a *meteor shower*. The point in the sky from which the meteors appear to emanate is known as the *radiant*. Most showers are known by the constellation in which the radiant is situated.

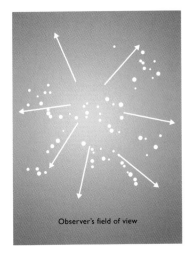

Observer's field of view

When watching meteors for a co-ordinated meteor programme, observers generally note the time, seeing conditions, cloud cover, their own location, the time and brightness of each meteor, and whether it was from the main meteor stream. It is also worth noting details of persistent afterglows (trains) and fireballs, and making counts of how many meteors appear in a given period.

COMETS

Comets are small bodies in orbit about the Sun. Consisting of frozen gases and dust, they are often known as 'dirty snowballs'. When their orbits bring them close to the Sun, the ices evaporate and dramatic tails of gas and dust can sometimes be seen.

A number of comets move round the Sun in fairly small, elliptical orbits in periods of a few years; others have much longer periods. Most really brilliant comets have orbital periods of several thousands or even millions of years. The exception is Comet Halley, a bright comet with a period of about 76 years. It was last seen with the naked eye in 1986.

Binoculars and wide-field telescopes provide the best views of comet tails. Larger telescopes with a high magnification are necessary to observe fine detail in the gaseous head (*coma*). Most comets are discovered with professional instruments, but a few are still found by experienced amateur astronomers.

We are expecting three comets visible in binoculars or a small telescope in 2017: Comet Encke in February; and then Comet Tuttle–Giacobini–Kresak in April/May, and Comet Johnson from April to June. Though astronomers don't currently know of any comets that will reach naked-eye brightness this year, a brilliant new comet could always put in a surprise appearance.

Deep-sky objects are 'fuzzy patches' that lie outside the Solar System. They include star clusters, nebulae and galaxies. To observe the majority of deep-sky objects you will need binoculars or a telescope, but there are also some beautiful naked-eye objects, notably the Pleiades and the Orion Nebula.

The faintest object that an instrument can see is its *limiting magnitude*. The table gives a rough guide, for good seeing conditions, for a variety of small- to medium-sized telescopes.

We have provided a selection of recommended deep-sky targets, together with their magnitudes. Some are described in more detail in our monthly 'Object' features. Look on the appropriate month's map to find which constellations are on view, and then choose your objects using the list below. We have provided celestial coordinates for readers with detailed star maps or Go To telescopes. The suggested times of year for viewing are when the constellation is highest in the sky in the late evening.

Limiting magnitude for small to medium telescopes	
Aperture (mm)	Limiting magnitude
50	+11.2
60	+11.6
70	+11.9
80	+12.2
100	+12.7
125	+13.2
150	+13.6

RECOMMENDED DEEP-SKY OBJECTS

Andromeda – autumn and early winter

M31 (NGC 224)
Andromeda Galaxy
3rd-magnitude spiral galaxy
RA 00h 42.7m Dec +41° 16'

M32 (NGC 221)
8th-magnitude elliptical galaxy, a companion to M31
RA 00h 42.7m Dec +40° 52'

M110 (NGC 205)
8th-magnitude elliptical galaxy
RA 00h 40.4m Dec +41° 41'

NGC 7662
Blue Snowball
8th-magnitude planetary nebula
RA 23h 25.9m Dec +42° 33'

Aquarius – late autumn and early winter

M2 (NGC 7089)
6th-magnitude globular cluster
RA 21h 33.5m Dec –00° 49'

M72 (NGC 6981)
9th-magnitude globular cluster
RA 20h 53.5m Dec –12° 32'

NGC 7293
Helix Nebula
7th-magnitude planetary nebula
RA 22h 29.6m Dec –20° 48'

NGC 7009
Saturn Nebula
8th-magnitude planetary nebula
RA 21h 04.2m Dec –11° 22'

Aries – early winter

NGC 772
10th-magnitude spiral galaxy
RA 01h 59.3m Dec +19° 01'

Auriga – winter

M36 (NGC 1960)
6th-magnitude open cluster
RA 05h 36.1m Dec +34° 08'

M37 (NGC 2099)
6th-magnitude open cluster
RA 05h 52.4m Dec +32° 33'

M38 (NGC 1912)
6th-magnitude open cluster
RA 05h 28.7m Dec +35° 50'

Cancer – late winter to early spring

M44 (NGC 2632)
Praesepe or Beehive
3rd-magnitude open cluster
RA 08h 40.1m Dec +19° 59'

M67 (NGC 2682)
7th-magnitude open cluster
RA 08h 50.4m Dec +11° 49'

Canes Venatici – visible all year

M3 (NGC 5272)
6th-magnitude globular cluster
RA 13h 42.2m Dec +28° 23'

M51 (NGC 5194/5)
Whirlpool Galaxy
8th-magnitude spiral galaxy
RA 13h 29.9m Dec +47° 12'

M63 (NGC 5055)
9th-magnitude spiral galaxy
RA 13h 15.8m Dec +42° 02'

M94 (NGC 4736)
8th-magnitude spiral galaxy
RA 12h 50.9m Dec +41° 07'

M106 (NGC4258)
8th-magnitude spiral galaxy
RA 12h 19.0m Dec +47° 18'

Canis Major – late winter

M41 (NGC 2287)
4th-magnitude open cluster
RA 06h 47.0m Dec –20° 44'

Capricornus – late summer and early autumn

M30 (NGC 7099)
7th-magnitude globular cluster
RA 21h 40.4m Dec –23° 11'

Cassiopeia – visible all year

M52 (NGC 7654)
6th-magnitude open cluster
RA 23h 24.2m Dec +61° 35'

M103 (NGC 581)
7th-magnitude open cluster
RA 01h 33.2m Dec +60° 42'

NGC 225
7th-magnitude open cluster
RA 00h 43.4m Dec +61 47'

NGC 457
6th-magnitude open cluster
RA 01h 19.1m Dec +58° 20'

NGC 663
Good binocular open cluster
RA 01h 46.0m Dec +61° 15'

Cepheus – visible all year

Delta Cephei
Variable star, varying between +3.5 and +4.4 with a period of 5.37 days. It has a magnitude +6.3 companion and they make an attractive pair for small telescopes or binoculars.

Cetus – late autumn

Mira (omicron Ceti)
Irregular variable star with a period of roughly 330 days and a range between +2.0 and +10.1.

M77 (NGC 1068)
9th-magnitude spiral galaxy
RA 02h 42.7m Dec –00° 01'

Coma Berenices – spring

M53 (NGC 5024)
8th-magnitude globular cluster
RA 13h 12.9m Dec +18° 10'

M64 (NGC 4286)
Black Eye Galaxy
8th-magnitude spiral galaxy with a prominent dust lane that is visible in larger telescopes.
RA 12h 56.7m Dec +21° 41'

M85 (NGC 4382)
9th-magnitude elliptical galaxy
RA 12h 25.4m Dec +18° 11'

M88 (NGC 4501)
10th-magnitude spiral galaxy
RA 12h 32.0m Dec.+14° 25'

M91 (NGC 4548)
10th-magnitude spiral galaxy
RA 12h 35.4m Dec +14° 30'

M98 (NGC 4192)
10th-magnitude spiral galaxy
RA 12h 13.8m Dec +14° 54'

M99 (NGC 4254)
10th-magnitude spiral galaxy
RA 12h 18.8m Dec +14° 25'

M100 (NGC 4321)
9th-magnitude spiral galaxy
RA 12h 22.9m Dec +15° 49'

NGC 4565
10th-magnitude spiral galaxy
RA 12h 36.3m Dec +25° 59'

Cygnus – late summer and autumn

Cygnus Rift
Dark cloud just south of Deneb that appears to split the Milky Way in two.

NGC 7000
North America Nebula
A bright nebula against the background of the Milky Way, visible with binoculars under dark skies.
RA 20h 58.8m Dec +44° 20'

NGC 6992
Veil Nebula (part)
Supernova remnant, visible with binoculars under dark skies.
RA 20h 56.8m Dec +31 28'

M29 (NGC 6913)
7th-magnitude open cluster
RA 20h 23.9m Dec +36° 32'

M39 (NGC 7092)
Large 5th-magnitude open cluster
RA 21h 32.2m Dec +48° 26'

NGC 6826
Blinking Planetary
9th-magnitude planetary nebula
RA 19 44.8m Dec +50° 31'

Delphinus – late summer

NGC 6934
9th-magnitude globular cluster
RA 20h 34.2m Dec +07° 24'

Draco – midsummer

NGC 6543
9th-magnitude planetary nebula
RA 17h 58.6m Dec +66° 38'

Gemini – winter

M35 (NGC 2168)
5th-magnitude open cluster
RA 06h 08.9m Dec +24° 20'

NGC 2392
Eskimo Nebula
8–10th-magnitude planetary nebula
RA 07h 29.2m Dec +20° 55'

Hercules – early summer

M13 (NGC 6205)
6th-magnitude globular cluster
RA 16h 41.7m Dec +36° 28'

M92 (NGC 6341)
6th-magnitude globular cluster
RA 17h 17.1m Dec +43° 08'

NGC 6210
9th-magnitude planetary nebula
RA 16h 44.5m Dec +23 49'

Hydra – early spring

M48 (NGC 2548)
6th-magnitude open cluster
RA 08h 13.8m Dec −05° 48'

M68 (NGC 4590)
8th-magnitude globular cluster
RA 12h 39.5m Dec −26° 45'

M83 (NGC 5236)
8th-magnitude spiral galaxy
RA 13h 37.0m Dec −29° 52'

NGC 3242
Ghost of Jupiter
9th-magnitude planetary nebula
RA 10h 24.8m Dec −18° 38'

Leo – spring

M65 (NGC 3623)
9th-magnitude spiral galaxy
RA 11h 18.9m Dec +13° 05'

M66 (NGC 3627)
9th-magnitude spiral galaxy
RA 11h 20.2m Dec +12° 59'

M95 (NGC 3351)
10th-magnitude spiral galaxy
RA 10h 44.0m Dec +11° 42'

M96 (NGC 3368)
9th-magnitude spiral galaxy
RA 10h 46.8m Dec +11° 49'

M105 (NGC 3379)
9th-magnitude elliptical galaxy
RA 10h 47.8m Dec +12° 35'

Lepus – winter

M79 (NGC 1904)
8th-magnitude globular cluster
RA 05h 24.5m Dec −24° 33'

Lyra – spring

M56 (NGC 6779)
8th-magnitude globular cluster
RA 19h 16.6m Dec +30° 11'

M57 (NGC 6720)
Ring Nebula
9th-magnitude planetary nebula
RA 18h 53.6m Dec +33° 02'

Monoceros – winter

M50 (NGC 2323)
6th-magnitude open cluster
RA 07h 03.2m Dec −08° 20'

NGC 2244
Open cluster surrounded by the faint Rosette Nebula, NGC 2237. Visible in binoculars.
RA 06h 32.4m Dec +04° 52'

Ophiuchus – summer

M9 (NGC 6333)
8th-magnitude globular cluster
RA 17h 19.2m Dec −18° 31'

M10 (NGC 6254)
7th-magnitude globular cluster
RA 16h 57.1m Dec −04° 06'

M12 (NCG 6218)
7th-magnitude globular cluster
RA 16h 47.2m Dec −01° 57'

M14 (NGC 6402)
8th-magnitude globular cluster
RA 17h 37.6m Dec −03° 15'

M19 (NGC 6273)
7th-magnitude globular cluster
RA 17h 02.6m Dec −26° 16'

M62 (NGC 6266)
7th-magnitude globular cluster
RA 17h 01.2m Dec −30° 07'

M107 (NGC 6171)
8th-magnitude globular cluster
RA 16h 32.5m Dec −13° 03'

Orion – winter

M42 (NGC 1976)
Orion Nebula
4th-magnitude nebula
RA 05h 35.4m Dec −05° 27'

M43 (NGC 1982)
5th-magnitude nebula
RA 05h 35.6m Dec −05° 16'

M78 (NGC 2068)
8th-magnitude nebula
RA 05h 46.7m Dec +00° 03'

Pegasus – autumn

M15 (NGC 7078)
6th-magnitude globular cluster
RA 21h 30.0m Dec +12° 10'

Perseus – autumn to winter

M34 (NGC 1039)
5th-magnitude open cluster
RA 02h 42.0m Dec +42° 47'

M76 (NGC 650/1)
Little Dumbbell
11th-magnitude planetary nebula
RA 01h 42.4m Dec +51° 34'

NGC 869/884
Double Cluster
Pair of open star clusters
RA 02h 19.0m Dec +57° 09'
RA 02h 22.4m Dec +57° 07'

Pisces – autumn

M74 (NGC 628)
9th-magnitude spiral galaxy
RA 01h 36.7m Dec +15° 47'

Puppis – late winter

M46 (NGC 2437)
6th-magnitude open cluster
RA 07h 41.8m Dec –14° 49'

M47 (NGC 2422)
4th-magnitude open cluster
RA 07h 36.6m Dec –14° 30'

M93 (NGC 2447)
6th-magnitude open cluster
RA 07h 44.6m Dec –23° 52'

Sagitta – late summer

M71 (NGC 6838)
8th-magnitude globular cluster
RA 19h 53.8m Dec +18° 47'

Sagittarius – summer

M8 (NGC 6523)
Lagoon Nebula
6th-magnitude nebula
RA 18h 03.8m Dec –24° 23'

M17 (NGC 6618)
Omega Nebula
6th-magnitude nebula
RA 18h 20.8m Dec –16° 11'

M18 (NGC 6613)
7th-magnitude open cluster
RA 18h 19.9m Dec –17 08'

M20 (NGC 6514)
Trifid Nebula
9th-magnitude nebula
RA 18h 02.3m Dec –23° 02'

M21 (NGC 6531)
6th-magnitude open cluster
RA 18h 04.6m Dec –22° 30'

M22 (NGC 6656)
5th-magnitude globular cluster
RA 18h 36.4m Dec –23° 54'

M23 (NGC 6494)
5th-magnitude open cluster
RA 17h 56.8m Dec –19° 01'

M24 (NGC 6603)
5th-magnitude open cluster
RA 18h 16.9m Dec –18° 29'

M25 (IC 4725)
5th-magnitude open cluster
RA 18h 31.6m Dec –19° 15'

M28 (NGC 6626)
7th-magnitude globular cluster
RA 18h 24.5m Dec –24° 52'

M54 (NGC 6715)
8th-magnitude globular cluster
RA 18h 55.1m Dec –30° 29'

M55 (NGC 6809)
7th-magnitude globular cluster
RA 19h 40.0m Dec –30° 58'

M69 (NGC 6637)
8th-magnitude globular cluster
RA 18h 31.4m Dec –32° 21'

M70 (NGC 6681)
8th-magnitude globular cluster
RA 18h 43.2m Dec –32° 18'

M75 (NGC 6864)
9th-magnitude globular cluster
RA 20h 06.1m Dec –21° 55'

Scorpius (northern part) – midsummer

M4 (NGC 6121)
6th-magnitude globular cluster
RA 16h 23.6m Dec –26° 32'

M7 (NGC 6475)
3rd-magnitude open cluster
RA 17h 53.9m Dec –34° 49'

M80 (NGC 6093)
7th-magnitude globular cluster
RA 16h 17.0m Dec –22° 59'

Scutum – mid to late summer

M11 (NGC 6705)
Wild Duck Cluster
6th-magnitude open cluster
RA 18h 51.1m Dec –06° 16'

M26 (NGC 6694)
8th-magnitude open cluster
RA 18h 45.2m Dec –09° 24'

Serpens – summer

M5 (NGC 5904)
6th-magnitude globular cluster
RA 15h 18.6m Dec +02° 05'

M16 (NGC 6611)
6th-magnitude open cluster,
surrounded by the Eagle Nebula.
RA 18h 18.8m Dec –13° 47'

Taurus – winter

M1 (NGC 1952)
Crab Nebula
8th-magnitude supernova remnant
RA 05h 34.5m Dec +22° 00'

M45
Pleiades
1st-magnitude open cluster,
an excellent binocular object.
RA 03h 47.0m Dec +24° 07'

Triangulum – autumn

M33 (NGC 598)
6th-magnitude spiral galaxy
RA 01h 33.9m Dec +30° 39'

Ursa Major – all year

M81 (NGC 3031)
7th-magnitude spiral galaxy
RA 09h 55.6m Dec +69° 04'

M82 (NGC 3034)
8th-magnitude starburst galaxy
RA 09h 55.8m Dec +69° 41'

M97 (NGC 3587)
Owl Nebula
12th-magnitude planetary nebula
RA 11h 14.8m Dec +55° 01'

M101 (NGC 5457)
8th-magnitude spiral galaxy
RA 14h 03.2m Dec +54° 21'

M108 (NGC 3556)
10th-magnitude spiral galaxy
RA 11h 11.5m Dec +55° 40'

M109 (NGC 3992)
10th-magnitude spiral galaxy
RA 11h 57.6m Dec +53° 23'

Virgo – spring

M49 (NGC 4472)
8th-magnitude elliptical galaxy
RA 12h 29.8m Dec +08° 00'

M58 (NGC 4579)
10th-magnitude spiral galaxy
RA 12h 37.7m Dec +11° 49'

M59 (NGC 4621)
10th-magnitude elliptical galaxy
RA 12h 42.0m Dec +11° 39'

M60 (NGC 4649)
9th-magnitude elliptical galaxy
RA 12h 43.7m Dec +11° 33'

M61 (NGC 4303)
10th-magnitude spiral galaxy
RA 12h 21.9m Dec +04° 28'

M84 (NGC 4374)
9th-magnitude elliptical galaxy
RA 12h 25.1m Dec +12° 53'

M86 (NGC 4406)
9th-magnitude elliptical galaxy
RA 12h 26.2m Dec +12° 57'

M87 (NGC 4486)
9th-magnitude elliptical galaxy
RA 12h 30.8m Dec +12° 24'

M89 (NGC 4552)
10th-magnitude elliptical galaxy
RA 12h 35.7m Dec +12° 33'

M90 (NGC 4569)
9th-magnitude spiral galaxy
RA 12h 36.8m Dec +13° 10'

M104 (NGC 4594)
Sombrero Galaxy
Almost edge-on 8th-magnitude
spiral galaxy.
RA 12h 40.0m Dec –11° 37'

Vulpecula – late summer and autumn

M27 (NGC 6853)
Dumbbell Nebula
8th-magnitude planetary nebula
RA 19h 59.6m Dec +22° 43'

OBSERVING ON HOLIDAY

Holidays give many people their only opportunity to do some proper astronomy in much darker skies than they normally get, and maybe see the Milky Way and other deep-sky sights for the first time. But for the really dedicated it can also mean finding very stable observing conditions for planetary imaging. Either way, the big problem is what instruments you should take with you, particularly if you're planning to fly and portability is important.

This becomes obvious when you travel by air, with a typical 20 kg baggage limit. Most portable instruments weigh a large fraction of this – the classic Meade ETX-90, for example, weighs in at 8.6 kg including the tripod. So while such an instrument is very versatile, you can find that either you have to compromise on whatever else you take with you, or pay considerable baggage excess fees. If you are planning to observe a specific astronomical event, such as a total solar eclipse, you'll probably want to make sure you have the right instrument to observe it so won't mind this. But for basic leisure observing, the telescope may have to take its place with other holiday essentials.

Many airlines will allow several people travelling on a single booking to pool their baggage allowance, though not if they are all booked separately, so find out in good time what you are actually allowed. There is also the carry-on baggage allowance, which must conform with size rules, and is usually around 7 kg, but you are usually limited to one or two items only.

PLANETARY OBSERVING

If you are keen on the Sun, Moon and planets, for most purposes just being abroad is no better than being at home. Light pollution is not usually a problem for planetary work, so there's little benefit in having a dark sky. But getting good seeing – that is, freedom from atmospheric turbulence – is the planetary observer's dream. While high mountain sites are generally reckoned to be ideal, steady conditions can sometimes also be found at sea level, particularly away from the jet stream. This high-level air stream snakes across the mid latitudes, and results in very poor seeing when it is present. It is usually absent from the tropics.

Some keen UK planetary observers take telescopes up to 14 inches (355 mm) in aperture to locations such as Barbados and get superb images. As well as the lack of a jet stream, the planets are very much higher in the sky, and the small temperature difference between day and night also helps to give good seeing. But you need to be dedicated to undertake such an expedition, and most of us will just take our chances with the British conditions, which can still give good imaging conditions from time to time.

DARK-SKY TARGETS

For most other types of observing, dark skies are the holy grail. Views of deep-sky objects such as nebulae and galaxies, comets and aurorae are transformed in a dark sky. Of course, the largest instruments generally give the best views, but most people will be happy with whatever views they can get. Binoculars come into their own, and even quite small glasses will show you objects that are hard to see with larger apertures from home. But these tantalizing glimpses won't always be enough, so what are the options for taking larger apertures on holiday without too much trouble?

The good news is that when the skies are really dark you can get away with much smaller telescopes than when you have poorer skies. Apertures of about 70 mm or 80 mm can give excellent views of many deep-sky objects. A short-focus refractor is not too heavy, and can give enough magnification to reveal useful detail on the smaller objects such as galaxies. And the darker skies allow you to see more than just the bright central regions, so the objects themselves appear larger.

One such instrument is the Celestron Travel Scope 70, which comes with its own small photographic tripod with a total weight of 1.5 kg. It has two eyepieces, giving magnifications of 20× and 40×, ideally suited to sweeping along the rich star fields of the Milky Way in dark skies. An extra eyepiece will give you more magnification, within reason.

A larger instrument is the Sky-Watcher Startravel, an 80 mm f/5 refractor. This comes in two forms – either as tube only (known in the trade as OTA, for optical tube assembly), or together with a tabletop German-type equatorial mount. Both cost around £100, with the OTA alone being slightly cheaper but instead of the German mount it has a reasonable optical finderscope, which is useful when locating faint objects. The German mount version can be motorized as an optional extra, so objects, once found, will stay in the field of view. Both come with eyepieces giving a magnification range from 16× to 80×.

However, the drawback with a German mount is that it needs a counterweight, which makes a lot of difference where weight is concerned. The OTA alone weighs 1.1 kg (plus eyepieces and finder), which won't make too much of a dent in your 20 kg airline allowance. But the German mount with counterweight weighs 3.74 kg. The tube alone will need a standard photographic tripod, typically weighing about 1½–2 kg, though much less easy to use when finding objects as it usually lacks the 'slow motion' knobs of the mount.

An alternative to the Startravel 80 refractor is the more expensive Skymax-90 Maksutov reflector, weighing 1.43 kg

▲ The Sky-Watcher Startravel 80 refractor on EQ1 tabletop equatorial mount. Like all equatorial mounts, this must be aligned correctly with the Earth's axis so that you can track objects using one slow-motion knob only, or using an optional battery-driven motor drive. Notice that there is a rather heavy counterweight which adds to the overall weight of the instrument. The unit has a red-dot finder.

plus accessories. This offers higher magnifications using the same eyepieces, with the range being 50× to 250×. It's a matter of preference whether you want higher powers for detailed viewing or the wider field of view offered by the refractor's lower power.

A very popular instrument with similar capabilities is the Meade ETX-90, now the only Maksutov left in a range that once offered 105 mm and 125 mm alternatives. Here we are getting into the realm of Go To instruments – that is, once set up, they will drive to any chosen object automatically. The telescope is integrated into its Go To mount and weighs 3.5 kg, but you really need the tripod as well, making a total of 8.5 kg.

Sky-Watcher have their own range of OTAs which will fit on to their Synscan Go To mount. Their Skymax 102 mm Maksutov, for example, weighs 2 kg and the mount and tripod add up to just under 4 kg. With a retracted length of 69 cm, even the tripod could fit into a suitcase.

▲ *The classic Meade ETX-90 is described as a 'portable observatory' and comes complete with a hard case. The mount can be tilted to your latitude so that it functions as an equatorial mount, and the unit is fully motorized with a Go To handset.*

▶ *The iOptron SkyTracker ready for use at the Kielder Forest Star Party in Northumberland with a DSLR and telephoto lens for long-exposure sky photography. It has its own polar tilt wedge and integrated pole-finding scope, but the ball-and-socket camera mount is an additional purchase.*

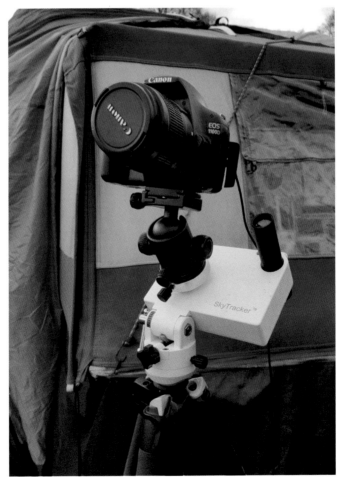

PHOTOGRAPHY

You can get some spectacular sky photos from dark-sky sites, but for the best results you need to be able to track the sky so as to avoid star trails during long exposures. The motorized Sky-Watcher equatorial mount will help, though it is not designed to take a camera directly, but another solution is to buy a purpose-designed, battery-driven sky tracker which will fit on to a photographic tripod. There are now several units on the market, such as the Vixen Polarie, the iOptron SkyTracker, the Sky-Watcher Star Adventurer and the Astro-trac, each costing between about £200 and £400. As with equatorial mounts, they must be aligned on the celestial pole before use, either by tilting them on a photographic tripod or using an angled block termed a wedge.

Typically they weigh about 1 kg and will take a payload of between 2 kg and 5 kg depending on the model, which is enough for a DSLR camera and telephoto lens. With some, such as the Sky-Watcher Star Adventurer, you could even attach a small telescope, so you could use it as a driven telescope mount. You'll need a fairly sturdy tripod – use a cheap plastic job and therein lies madness.

You will also need a ball and socket or similar so as to be able to point the camera at any part of the sky, and maybe a polefinder scope if it isn't included. But armed with such a set-up you can both observe and take good images of the Milky Way or individual objects without having to approach the check-in desk with trepidation in case they spot that your baggage is drastically overweight!

WHAT WILL THEY COST?

All equipment mentioned here is widely available from specialist retailers. Guide prices are given for basic equipment and extras may be needed.

Celestron Travel Scope 70
= £80

Sky-Watcher Startravel-80 Tabletop
= £119

Sky-Watcher Startravel-80 OTA only
= £109

Sky-Watcher Skymax-90T OTA only
= £139

Sky-Watcher Skymax-102 Maksutov Synscan AZ Go To
= £379 (including tripod)

Meade ETX-90 90mm Maksutov Portable Observatory w/Hard Case
= £450 (including tripod)

Sky-Watcher Star Adventurer
= £219

Vixen Polarie Star Tracker
= £299

iOptron SkyTracker
= £299

Astrotrac TT320X-AG
= £395

◄ *A holiday target – the Large Magellanic Cloud, photographed using a portable tracking mount in dark skies near Benalla, Victoria, Australia. The author used a Canon 70D camera with telephoto lens and a total exposure time of 3 minutes.*